The Atmosphere and Ocean:
A Physical Introduction

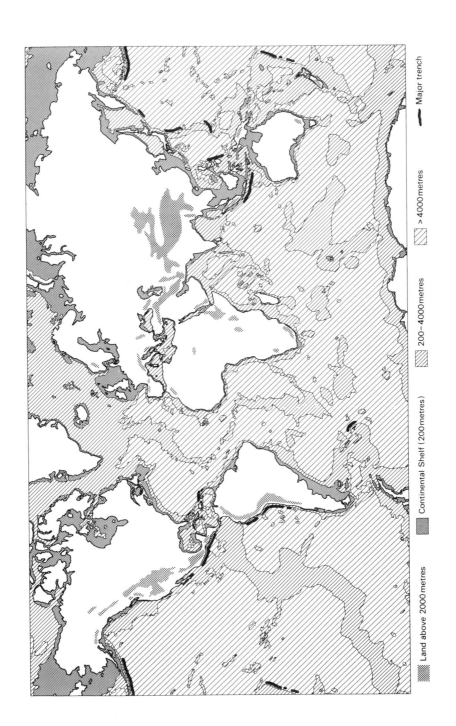

Land above 2000 metres   Continental Shelf (200 metres)   200–4000 metres   >4000 metres   Major trench

Frontispiece.   Major topographic features of the ocean and the continents.

The Atmosphere and Ocean:
A Physical Introduction

Neil Wells

Department of Oceanography, University of Southampton

Taylor & Francis
London and Philadelphia
1986

UK   Taylor & Francis Ltd, 4 John St, London WC1N 2ET

USA   Taylor & Francis Inc., 242 Cherry St, Philadelphia,
            PA 19106-1906

British Library Catologuing in Publication Data

Wells, Neil
    The atmosphere and ocean: a physical introduction.
    1. Oceanography
    I. Title
    551.46   GC11.2

    ISBN 0-85066-259-1 (pbk)
    ISBN 0-85066-348-2 (hbk)

Library of Congress Cataloging in Publication Data

Wells, Neil.
    The atmosphere and ocean.
    Bibliography: p.
    Includes index.
    1. Atmospheric physics. 2. Oceanography. I. Title.
    QC861.2.W45   1986   551.5   85-26224
    ISBN 0-85066-259-1 (pbk)
    ISBN 0-85066-348-2 (hbk)

The front cover photograph shows a waterspout in the North-East Atlantic in March 1984. Photograph courtesy of Karen Thomas, University of Southampton.

Typeset by Mathematical Composition Setters Ltd, 7 Ivy Street, Salisbury, UK
Printed in Great Britain by Taylor & Francis (Printers) Ltd, Basingstoke, Hants.

Preface

There are many texts devoted to either meteorology or oceanography, but only a few have attempted to bring both disciplines together. The ocean and atmosphere are constantly exchanging kinetic energy, heat, momentum and matter, and both systems are ultimately powered by solar radiation; therefore they cannot be treated solely in isolation from one another. The purpose of the approach adopted has been twofold. First, to bring together a description of the atmosphere and ocean, with emphasis on their physical properties and interdependence. Second, to demonstrate that atmosphere and ocean are both fluids subject to the influence of the Earth's rotation and therefore that they have a common dynamical basis. Whilst emphasizing the similar attributes of the two systems, and their interactions, it has not been the intention of the author to ignore the fundamental differences between the two environments and, where necessary, separate approaches have been utilized.

The book is broadly based on courses given to first and second year undergraduate science students at Southampton University over the past five years. The descriptive elements of the book should be suitable for an introductory course, whilst the dynamical treatment (Chapters 7, 8 and 9) would be more suitable for second-year students. The overall aim of the book has not been to give a detailed grounding of the dynamics of the two systems, which is well taken care of in more specific texts, but it is hoped that it will provide students with some insights into the physical processes in the atmosphere and ocean, as well as giving an indication of how the subject is currently developing. Two major experiments in the World Climate programme, TOGA (Tropical Ocean–Global Atmosphere) and WOCE (World Ocean Circulation Experiment), which are planned for the next decade, underline the need for students to be introduced to both systems early in their careers.

It is a pleasure to acknowledge my colleagues and students in the Department of Oceanography, who have given encouragement during the writing of this book as well as providing many inadvertent insights into the complexity of the atmosphere–ocean system. Special thanks go to Jean Watson for typing the manuscript (often at the weekends), to John Cross and Jenny Mallinson for preparing many of the diagrams, and to Simon Boxall, Ian Robinson and Chris Mead for proof reading the final manuscript. Finally, I would like to thank my wife for her considerable assistance at all stages in the preparation of this book, whilst attending to and caring for a demanding small family.

Neil Wells
Southampton
September 1985

Contents

1

The Earth within the solar system

1.1. The Sun and its constancy

Any account of the Earth's atmosphere and ocean cannot be regarded as complete without a discussion of the Sun, the solar system and the place of the Earth within this system. The Sun supplies the energy absorbed by the Earth's atmosphere–ocean system. Some of its energy is converted directly into thermal energy, which drives the atmospheric circulation. A small portion of this energy appears as the kinetic energy of the winds which, in turn, drives the ocean circulation. Some of the intercepted solar energy is transformed by photosynthesis into biomass, a large proportion of which is ultimately converted into heat energy by chemical oxidation within the bodies of animals and by the decomposition and burning of vegetable matter. A very small proportion of the photosynthetic process produces organic sediments, which may be eventually transformed into fossil fuels. It is estimated that the solar radiation intercepted by the Earth in 10 days is equivalent to the heat which would be released by the combustion of all known reserves of fossil fuels on Earth. The Sun is therefore of fundamental importance in the understanding of the uniqueness of the Earth.

The Sun is a main sequence star in the middle stages of its life and was formed $4 \cdot 6 \times 10^9$ years ago. It is composed mainly of hydrogen (74% by mass) and helium (25% by mass); the remaining 1% of the Sun's mass comprises the elements oxygen, nitrogen, carbon, silicon, iron, magnesium and calcium. The emitted energy of the Sun is $3 \cdot 8 \times 10^{26}$ W and this energy emission arises from the thermonuclear fusion of hydrogen into helium at temperatures around $1 \cdot 5 \times 10^6$ K in the core of the Sun, as given by equation 1.1:

$$4^1_1H \rightarrow ^4_2He + 2\beta + \text{energy} (26 \cdot 7 \text{ MeV}) \tag{1.1}$$

In the core, the dominant constituent is helium (65% by mass) and the hydrogen content is reduced to 35% by mass as a direct result of its consumption in the fusion reactions. It is estimated that the remaining hydrogen in the Sun's core is sufficient to maintain the Sun at its present luminosity and size for a further 4×10^9 years. At this stage it is expected that the Sun will expand into a red giant and engulf all of the inner planets of the solar system (i.e. Mercury, Venus, Earth and Mars).

There exists a high-pressure gradient between the core of the Sun and its perimeter and this is balanced by the gravitational attraction of the mass of the Sun. In the core, the energy released by the thermonuclear reaction is transported by energetic photons, but, because of the strong absorption by the peripheral gases, most of these photons do not penetrate the surface. This absorption causes heating in the region outside the core. In contrast, the outer layers of the Sun are continually losing energy by radiative emission into space in all regions of the electromagnetic spectrum. This causes a large temperature gradient to develop between the surface and the inner region of the Sun. This large temperature gradient produces an unstable region and large-scale convection currents are set up which transfer heat to the surface of the Sun. These convection currents are visible as the fine grain structure, or granules, in high-resolution photographs of the surface of the Sun. It is thought that the convection currents have a three-tier structure within the Sun. The largest cells, 200×10^3 km in diameter, are close to the core. In the middle tier the convection cells are about 30×10^3 km in diameter, and at the surface they are 10^3 km. The latter cells have a depth of 2000 km. In each cell, hot gas is transported towards the cooler surface whilst the return flow transports cooler gas towards the interior.

Almost all of the solar radiation emitted into space, approximately $99 \cdot 9\%$, originates from the visible disc of the Sun, known as the photosphere. The photosphere is the region of the Sun where the density of the solar gas is sufficient to produce and emit a large number of photons, but where the density of the overlying layers of gas is insufficient to absorb the emitted photons. This region of the Sun has a thickness of approximately 500 km, but no sharp boundaries can be defined. The radiation spectrum of the Sun, when fitted to a blackbody curve, gives a blackbody† temperature of 6000 K, although the effective temperature, deduced from the total energy emitted by the Sun, i.e. its luminosity, gives a lower temperature of 5800 K.

The photosphere is not uniform in temperature. The lower regions of the photosphere have temperatures of 8000 K, whilst the outer regions have temperatures of 4000 K. Furthermore, the convection cells produce horizontal temperature variations of 100 K between the ascending and descending currents of solar gas. Larger convection cells also appear in

---

†A perfect emitter and absorber of radiation at all wavelengths is known as a blackbody.

the photosphere and they have diameters of 30,000 km. They appear to originate from the second tier of convection within the Sun. The appearance of sunspots give rise to horizontal variations of 2000 K within the photosphere. The inner regions of the sunspot have blackbody temperatures of 4000 K. Sunspots have diameters of 10,000–150,000 km and they may last for many weeks. It is thought that the 'sunspot' causes a localized suppression of convection and therefore leads to a reduction in the transfer of heat into the photosphere from the interior. However, although the sunspot features are dramatic, they occupy less than 1% of the Sun's disc, and therefore the effect on the luminosity of the Sun is small.

Beyond the photosphere lies the chromosphere where the temperature decreases to a minimum of 4000 K at 2000 km above the photosphere and then increases sharply to a temperature of 10^6 K at a height of 5000 km, in the region of the corona. However, because of the low density of the gas in this region, the radiation emitted from the chromosphere and the corona only amout to $0 \cdot 1\%$ of the total radiation from the Sun.

These different temperature zones of the Sun, can also be observed in the solar spectrum (figure 1.1). The visible and infra-red radiation emitted from the photosphere follow reasonably closely the blackbody

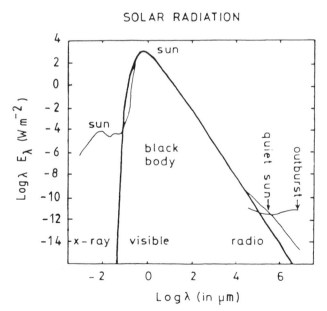

*Figure 1.1. Solar spectrum and blackbody curve: energy distribution of the Sun and a blackbody at 5800 K. (Reproduced from C. W. Allen (1958), Q. J. Roy. Met. Soc., **84**, 311, figure 3 with permission of the Royal Meteorological Society.)*

curve for a temperature of 6000 K. However, substantial deviations from the theoretical curve occur in the X-ray and radio wavebands, and lesser deviations occur in the ultraviolet spectrum. The high temperature of the corona is responsible for the intense X-ray band, whilst the high radio frequency energy is associated with the solar wind and solar activity. However, the energy in these wavebands is a negligible fraction of the total emitted energy and therefore these very variable regions of the spectrum have little direct influence on the total solar energy received on the Earth. The depletion of energy in the ultraviolet spectrum is the result of emission at a lower temperature than the photosphere and therefore probably originates in the temperature minimum of the lower chromosphere. Recent observations have shown that the ultraviolet energy is not constant and shows considerable variability in the short-wave part of the spectrum, amounting to 25% of its average value, at $0 \cdot 15 \ \mu m$, and 1% at $0 \cdot 23 \ \mu m$. Again the amount of energy in this band is relatively small and contributes less than $0 \cdot 1 \%$ to the total energy. Furthermore, satellite observations of the solar spectrum over a 9 month period have shown that the total solar output is constant to $0 \cdot 1 \%$, which is the present limit of experimental accuracy.

Thus it has been shown how constant the energy output of the Sun is, and it is now in order to discuss the variable amounts of solar radiation intercepted by the Earth as it orbits the Sun.

1.2. Orbital variations in solar radiation

Let S be the total solar output of radiation in all frequencies. At a distance, r, imagine a sphere of radius r on which the flux of radiation will be the same (assuming that the radiation from the Sun is equal in all directions). If the flux of radiation per unit area at distance r is given by $Q(r)$, then the total radiation on the imagined sphere is $4\pi r^2 Q(r)$.

In the absence of additional energy sources to the Sun:

$$S = 4\pi r^2 Q(r)$$

Rearranging:

$$Q(r) = \frac{S}{4\pi r^2} \qquad (1.2)$$

In practice, the solar radiation cannot be measured at the Sun but it can be measured by satellites above the Earth's atmosphere. Recent determinations of the flux of radiation per unit area, Q, give a value of 1360 W m^{-2}. Given that the Earth is approximately $1 \cdot 5 \times 10^{11}$ m away from the Sun, S can then be calculated to be $3 \cdot 8 \times 10^{26}$ W.

Table 1.1. Radiative properties of terrestrial planets. Q is the solar irradiance at distance r from the Sun, α is the the planetary albedo, T_e is the radiative equilibrium temperature and T_s is the surface temperature.

	r (10^9 m)	Q (W m^{-2})	α	T_e (K)	T_s (K)
Venus	108	2623	0·77	227	760
Earth	150	1360	0·30	255	288
Mars	228	589	0·24	211	230

Though S is the true solar constant, in meteorology Q is defined as the solar constant for the Earth. Table 1.1 shows values of the solar constant obtained for other planets in the solar system. It is noted that the dramatic changes in the solar constant between Earth and our nearest planetary neighbours, Mars and Venus, merely serve to highlight the uniqueness of the position of the Earth in the solar system. At the radius of Pluto (some 39 Earth–Sun distances) the flux of the radiation from the Sun is less than 1 W m^{-2}.

The radiation incident on a spherical planet is not equal to the solar constant of that planet. The planet intercepts a disc of radiation of area πa^2, where a is the planetary radius, whereas the surface area of the planet is $4\pi a^2$. Hence, the solar radiation per unit area on a spherical planet is

$$\frac{Q\pi a^2}{4\pi a^2} = \frac{Q}{4} \tag{1.3}$$

The the average radiation on the Earth's surface is 340 W m^{-2}. The above discussion assumes the total absence of an atmosphere and that the Earth is a perfect sphere, in a spherical orbit.

The three geometrical factors which determine the seasonal variation of solar radiation incident on the Earth are shown in figure 1.2. The Earth revolves about the Sun in an elliptical orbit, being closest to the Sun, i.e. at perihelion, about 3 January and farthest from the Sun, i.e. at aphelion, on 5 July. The eccentricity, represented by the symbol ϵ, of the present-day orbit is $0·017$. At aphelion it can be shown that

$$r_{\text{aphelion}} = (1 + \epsilon)\bar{r}$$

where $\bar{r}$ is the mean distance between the Earth and the Sun.

As the incident radiation is inversely proportional to the square of the Earth's distance from the Sun, it can also be shown that, at aphelion, the radiation received by the Earth will be $3·5\%$ less than the annual average. Furthermore, the total radiation incident on the Earth over the course of 1 year is independent

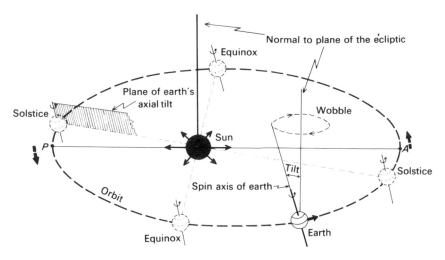

Figure 1.2. Geometry of the Earth–Sun system. The Earth's orbit, the large ellipse with major axis AP and the Sun at one focus, defines the plane of the elliptic. The plane of the Earth's axial tilt (shaded) is shown passing through the Sun corresponding to the time of the southern summer solstice. The Earth moves around its orbit in the direction of the solid arrow (period 1 year) while spinning about its axis in the direction shown by the thin curved arrows (period 1 day). The broken arrows shown opposite the points of aphelion (A) and perihelion (P) indicate the direction of the very slow rotation of the orbit. (Reproduced from Climate Change and Variability: A Southern Perspective, *edited by A. B. Pittock* et al., *p. 10, figure 2.1.1, with permission of Cambridge University Press.)*

of the eccentricity of the Earth's orbit. The long-term variation in eccentricity indicates changes of $0 \cdot 005 - 0 \cdot 060$ occurring with a period of 100,000 years. This would produce changes of the order of $\pm 10\%$ in the radiation incident at aphelion and perihelion.

However, the angle of tilt of the Earth with respect to the plane of rotation of the Earth's orbit, i.e. the obliquity of the ecliptic, is the dominant influence on the seasonal cycle of solar radiation (figure 1.3). It not only determines the march of the seasons but it is important in determining the latitudinal distribution of the climatic zones. At the present time this angle is $23 \cdot 5°$, which is similar to the obliquity of Mars. It can be shown that the amplitude of the seasonal variation in the solar radiation is directly proportional to the obliquity. If the obliquity were reduced, there would be a reduction in the amplitude of the seasonal solar radiation cycle, and if the obliquity were zero, the seasonal variation would depend solely on the eccentricty of the orbit.

The obliquity has varied between 22 and $24 \cdot 5°$ over the past 1 million years and this variation has had a period of 41,000 years. Such variations cannot influence the total radiation intercepted by the Earth but they will exert an influence on the seasonal cycle. The sensitivity of the global temperature to the seasonal cycle, which arises from ice, cloud and ocean feedback processes,

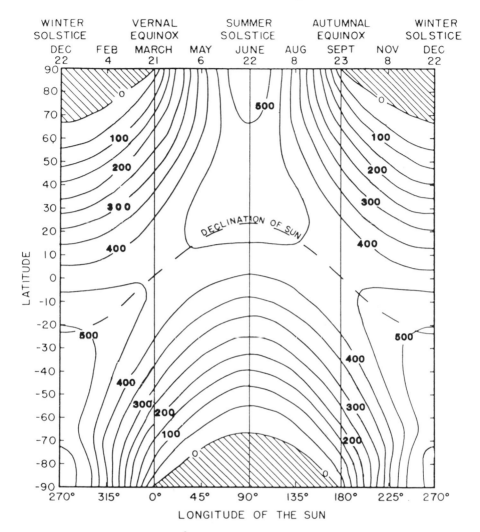

Figure 1.3. Solar radiation in W m⁻² arriving at the Earth's surface in the absence of an atmosphere. (Reproduced from Introduction to Theoretical Meteorology. *by S.L. Hess, p. 132, figure 9.1, with permission of Holt, Rinehart and Winston, New York.)*

indicates that these changes may be sufficient to initiate long-term changes in climate and, in the most extreme cases, ice ages.

The third parameter which affects the seasonal cycle is the longitude of the perihelion, measured relative to the vernal equinox. It precesses by 360° in a period of 21,000 years. Therefore, in about 10,500 years' time, the Earth will be closest to the Sun in July and farthest away in January. Current theories on the initiation of ice ages suggest that they develop in the Northern Hemisphere under orbital conditions which produce a relative decrease in solar radiation in the summer season (i.e. that the build-up of ice is associated with a decrease in melting in the summer). The optimum orbital conditions for this reduction in the summer solar radiation are:

(i) That aphelion occur during the Northern Hemisphere summer.
(ii) That the eccentricity be large, to maximize the Earth–Sun separation at aphelion.
(iii) That the obliquity be small.

These three factors would act to decrease the amplitude of the seasonal solar radiation cycle, and therefore reduce the solar radiation in summer at high latitudes in the Northern Hemisphere. However, all of these factors would tend to lead to an increase in solar radiation in the Northern Hemisphere winter. Comparison of the orbital parameters with long-term temperature records, deduced from the dating of deep-ocean sediment cores, indicate that there is reasonable statistical agreement with theory; but, because of the lack of detailed physical models of these long-term climate processes, there has been no definitive verification of these results.

1.3. *Radiative equilibrium temperature*

The inner planets of the solar system have one common attribute—the lack of a major source of internal energy. It is of course true that there is a steady geothermal flow of energy from the interior of the Earth as a result of radioactive decay in the Earth's core, but this energy flow is less than $0 \cdot 1 \text{ W m}^{-2}$ and can be compared with an average solar heat flux of 340 W m^{-2}. What, then, is the fate of this steady input of solar energy? On average, about 30% is scattered back into space, by clouds, snow, ice, atmospheric gases and aerosols and the land surface, leaving 70% available for the heating of the atmosphere, the ocean and the Earth's surface. However, the annual average of the Earth's temperature has not changed by more than 1 K over the past 100 years. Why, then, have the temperatures of the atmosphere and the ocean not increased? The observation that both the temperature of the atmosphere and the ocean have remained relatively constant implies that energy is being lost from the ocean, the atmosphere and the Earth's surface at approximately the same rate as it is being supplied by the Sun. The only significant mechanism by which this heat can be lost is by the emission of electromagnetic radiation from the Earth's atmosphere into space. If it is assumed that the Earth can be represented as a blackbody, then it is possible to apply Stefan's law to the electromagnetic emission of the Earth. Stefan's law states that the total emission of radiation from a blackbody over all wavelengths is proportional to the fourth power of the temperature, i.e.

$E = \sigma T^4$, where σ is the Stefan–Boltzmann constant. For the emission by the Earth of an amount of radiation equal to that received from the Sun there can be defined a radiative equilibrium temperature, T_e.

If $Q/4$ is the energy flux from the Sun and α is the fraction of solar radiation reflected back into space, then for the Earth's system in thermal equilibrium:

$$(Q/4)(1 - \alpha) = \sigma T_e^4 \qquad (1.4)$$

where $\sigma = 5 \cdot 67 \times 10^{-8} \, \text{W m}^{-2} \text{K}^{-4}$. By inserting $\alpha = 0 \cdot 3$ and $Q/4 = 340 \, \text{W m}^{-2}$, a radiative equilibrium temperature of 255 K, or $-18°\text{C}$, can be calculated.

This temperature is rather lower than might be expected. The mean surface temperature of the Earth is $15°\text{C}$, or approximately 288 K, and so it is clear that the radiative equilibrium temperature bears little direct relationship to the observed surface temperature. This is because most of the planetary radiation is emitted by the atmosphere, whilst only a small fraction originates from the surface of the Earth. The temperature of the atmosphere decreases by $6 \cdot 5 \, \text{K km}^{-1}$ from the surface to the tropopause, 10 km above the Earth's surface, and therefore the radiative equilibrium temperature corresponds to the temperature at a height of 5 km.

The Wien displacement equation enables the calculation of the wavelength of maximum radiation, λ_{max}, and it states that:

$$\lambda_{max} T = 2897 \, \mu\text{m K} \qquad (1.5)$$

If the radiative equilibrium temperature, $T_e = 255$ K, is substituted in this equation, then λ_{max} is calculated to be approximately 11 μm, which lies in the middle part of the infra-red spectrum. In this region of the electromagnetic spectrum, water and carbon dioxide have large absorption bands so that all substances containing water are particularly good absorbers of radiation in the middle infra-red. As far as emission is concerned, water has an emissivity of $0 \cdot 99$, and so clouds, which are composed mainly of water droplets, are good infra-red emitters. The prevalence of liquid water and water vapour in the Earth's atmosphere therefore leads to the strong absorption and re-emission of radiation. This ability of water to absorb and re-emit radiation back to the Earth's surface results in the higher observed mean surface temperature. If the atmosphere was transparent to the emitted planetary radiation, then the surface temperature would be close to the radiative equilibrium temperature. The temperature of the surface of a planet without an atmosphere, such as Mercury, would be observed to have a surface equilibrium temperature equal to that of the radiative equilibrium temperature. Table 1.1 shows the radiative equilibrium temperatures for Mars and Venus, as well as their surface temperatures. The surface temperatures are higher than the radiative temperatures. On Mars the atmospheric mass is smaller than that of the Earth by two orders of magnitude and therefore, though the carbon dioxide absorbs and re-emits planetary radiation back to the surface, a surface warming of only 19 K is produced. However, on Venus the surface temperature of 760 K is sufficient to melt lead in spite of the fact that the radiative equilibrium temperature is less than that of the Earth. This low radiative equilibrium temperature is caused by the reflection of 77% of the incident solar radiation back into space by the omnipresent clouds, which are not composed of water droplets as they are on Earth. Therefore, though the planet receives less net solar radiation than the Earth, it has a surface temperature 472 K higher. This is the result of the massive carbon dioxide atmosphere, which absorbs virtually all of the radiation emitted by the surface and re-emits it back to the surface. Furthermore, it is known that clouds of sulphuric acid also enhance this warming effect.

It is clear that, although the distance from the Sun determines the energy input into an atmosphere, the mass and the constituents of that atmosphere are also important factors in the determination of the surface temperature of the planet. On a speculative note, it may be seen that Mars would only require an atmosphere slightly more massive than that of the Earth to attain a mean surface temperature close to the freezing point of water.

1.4. *Thermal inertia of the atmosphere*

The thermal inertia of the atmosphere gives an indication of how quickly the atmospheric temperature would respond to variations in solar radiation, and it is therefore of importance in the understanding of climatic change. Consider an atmosphere having a mass M, per unit area, and a specific heat C_p. The thermal inertia of the atmosphere is MC_p.

Initially, the atmosphere is in thermal equilibrium and therefore the solar radiation absorbed by the atmosphere is in balance with the emission of long wavelength planetary radiation into space (equation 1.4).

Now consider the situation where heat is suddenly added to the atmosphere, for example as the result of the burning of fossil fuels or by a large thermonuclear explosion. How long would the atmosphere take to regain its former equilibrium? If all the heat were liberated at the same time, then the atmosphere's temperature would increase suddenly by ΔT. The planetary radiation emitted into space would now be $\sigma(T_e + \Delta T)^4$, and therefore more radiation would be emitted than received from the Sun. This deficit implies that the atmosphere would cool. The temperature change, dT/dt, is proportional to the net difference between the emitted planetary radiation and the solar radiation, and thus

$$MC_p \frac{dT}{dt} = -\sigma(T_e + \Delta T)^4 + \frac{Q}{4}(1 - \alpha)$$

Rearranging and substituting from equation 1.4:

$$MC_p \frac{dT}{dt} = -\sigma T_e^4 \left(1 + \frac{\Delta T}{T_e}\right)^4 + \sigma T_e^4$$

Expanding by the binomial expansion:

$$MC_p \frac{dT}{dt} = -\sigma T_e^4 \left(1 + \frac{4\Delta T}{T_e} + 6\left[\frac{\Delta T}{T_e}\right]^2 + 4\left[\frac{\Delta T}{T_e}\right]^3 + \left[\frac{\Delta T}{T_e}\right]^4\right) + \sigma T_e^4$$

Providing that $\Delta T/T_e \ll 1$, then the higher-order terms in the above expression can be neglected. Therefore, retaining only the first-order terms:

$$MC_p \frac{dT}{dt} = -\sigma T_e^4 \frac{4\Delta T}{T_e}$$

Since $T = T_e + \Delta T$, where T_e is independent of time. Then

$$\frac{dT}{dt} = \frac{d\Delta T}{dt}$$

Hence

$$\frac{d}{dt} \Delta T = -\left(\frac{4\sigma T_e^3}{MC_p}\right) \Delta T \tag{1.6}$$

The solution of the above equation is

$$\Delta T(t) = \Delta T_0 \exp(-t/\tau) \tag{1.7}$$

where ΔT_0 is the initial temperature perturbation at $t = 0$ and τ_R is known as the radiation relaxation time constant:

$$\tau_R = \frac{MC_p}{4\sigma T_e^3} \tag{1.8}$$

For $M = 10 \cdot 316 \times 10^3 \, \text{kg m}^{-2}$, $\sigma = 5 \cdot 67 \times 10^{-8} \, \text{W m}^{-2} \text{K}^{-4}$, $C_p = 1004 \, \text{J kg}^{-1} \text{K}^{-1}$ and $T_e = 255 \, \text{K}$, τ_R can be calculated to be $2 \cdot 7 \times 10^6$ s or

32 days. From equation 1.7

$$\Delta T = \Delta T_0\, e^{-1} \quad \text{at } t = \tau_R$$
$$\Delta T = \Delta T_0\, e^{-2} \quad \text{at } t = 2\tau_R$$
and
$$\Delta T = \Delta T_0\, e^{-3} \quad \text{at } t = 3\tau_R$$

For an initial temperature perturbation of 1 K, equivalent to the instantaneous burning of 240×10^9 t of coal, then after approximately 32 days, the temperature perturbation will be reduced to $0 \cdot 37$ K; after 64 days it will be $0 \cdot 13$ K and after 96 days it will be $0 \cdot 05$ K. For the Earth's atmosphere, therefore, the colossal excess of heat will be lost by planetary radiation to space within 100 days, and the Earth's radiative equilibrium will be restored.

Although this particular problem is hypothetical, the concept is a useful one for understanding the response of the atmosphere to changes in the radiative balance caused by the daily and seasonal variations in solar radiation. For instance, because the radiative time constant is much greater than 1 day, the lower part of the Earth's atmosphere (wherein is located the major part of the mass of the atmosphere) does not respond significantly to the daily variation in solar radiation. However, for seasonal variations, where the solar radiation changes are of a longer period than the radiative time constant, there will be a signficant response in the radiation temperature of the atmosphere.

Table 1.2 shows a comparison of the radiative time constant of Mars, Venus and the Earth. On Mars, because of the small thermal inertia, the diurnal variation of temperature at the surface is very large and as a result large thermal tides are set up in the atmosphere. This phenomenon is also found in the upper reaches of the Earth's atmosphere, where the atmospheric mass is small and the thermal capacity is reduced. However, on Venus the diurnal variation in temperature is only $1-2$ K, because of the very large radiative time constant, even though the Venusian day is 243 Earth days long!

Table 1.2. Radiative time constant, τ_R, for the terrestial planets. C_p is the specific heat at constant pressure, p is the surface pressure and g is the acceleration due to gravity.

	C_p ($J\ kg^{-1}\ K^{-1}$)	P (K Pa)	g ($m\ s^{-2}$)	τ_R (days)
Venus	830	9000	$8 \cdot 9$	3657
Earth	1004	100	$9 \cdot 8$	32
Mars	830	$0 \cdot 8$	$3 \cdot 7$	1

It must, of course, be remembered that the seasonal change in temperature on Earth is modified by the thermal capacity of the ocean. The thermal capacity of the atmosphere is equivalent to 2·6 m of sea water and therefore it can be seen that the thermal capacity of the ocean is equivalent to approximately 1600 atmospheres. However, during the seasonal cycle, only the upper 100 m of the ocean is affected significantly and therefore the thermal capacity is approximately equivalent to only 38 atmospheric masses, giving a thermal relaxation time of approximately 3·5 years. This simple sum suggests that the ocean is responsible for a major modification of the seasonal cycle of temperature in the atmosphere. An example of the oceanic influence is shown in figure 1.4. The flux of the long-wave radiation from the top of the atmosphere gives an indication of the seasonal change in temperature. In the Northern Hemisphere, where the largest ratio of continents to ocean (see figure 1.9) exists (some 70% at 60°N) there is a large change of approximately 40 K in the radiation temperature during the seasonal cycle, whilst in the

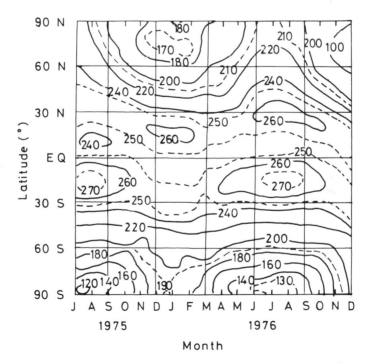

Figure 1.4. Seasonal variation of the zonal mean long-wave radiation flux, Wm^{-2}, measured by satellite from July 1975 to December 1976. A flux of 270 Wm^{-2} corresponds to a radiation temperature of 263 K, whilst a flux of 170 Wm^{-2} is equivalent to a temperature of 234 K. (Reproduced from H. Jacobowitz et al. (1979), J.A.S., 36, 506, figure 6 with permission of the American Meteorological Society.)

Southern Hemisphere between 30 and 60°S, where the continental area is less than 10%, there is only a small change in radiation temperature during the year. Over Antarctica, the seasonal change in temperature is large, because of the redistribution of heat by the atmospheric circulation systems. Even the most continental region benefits from an amelioration in its seasonal climate brought about by the ocean and even in Siberia the winters are not as severe as they would be if the ocean did not exist!

1.5. Albedo

The planetary albedo is the ratio of the reflected (scattered) solar radiation to the incident solar radiation, measured above the atmosphere. As this reflected radiation is lost immediately from the Earth–atmosphere system, the accurate determination of planetary albedo is important in the calculation of the amount of solar radiation absorbed by the system and therefore the climate of the Earth. On average, the global albedo is 30% ($\mp 2\%$). This value has been determined from a number of satellite measurements over a period of a few years. The global albedo is not constant but varies on both seasonal and interannual time scales by approximately 2%. Recent observations have shown that the albedo is a maximum in January and a minimum in July. If the cloud and surface distributions were similar in both hemispheres, then no annual variation would be expected. However, because of the large seasonal cycle in snow extent over the Northern Hemisphere continents, particularly over the Eurasian land mass, and the seasonal cycle of cloud in the Northern Hemisphere, the global albedo has a seasonal component.

The latitudinal variation in albedo is determined by:

(i) The elevation of the Sun.
(ii) The distribution and type of cloud.
(iii) The surface albedo.

At latitudes less than 30° the planetary albedo is relatively constant at 25%, but it then increases with latitude to reach a maximum value of 70% over the poles. However, because the polar regions occupy less than 8% of the Earth's surface, the value of the global albedo is weighted to the albedo of the middle and low latitudes. It is noted that the latitudinal distributions in the albedo in the Southern and Northern Hemispheres are similar, despite the differences in land and sea distribution between the two hemispheres. This indicates that the cloud distributions in both hemispheres have a dominant influence over the surface variations in albedo in determining the planetary albedo (see figure 1.5).

The long-term latitudinal variation in absorbed solar energy and the emitted planetary radiation is depicted in figure 1.6. The absorbed solar

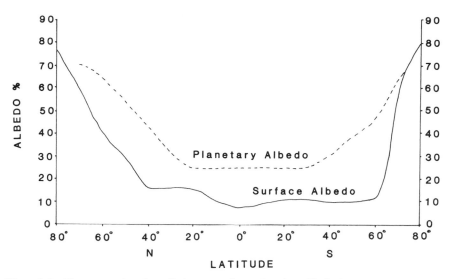

Figure 1.5. Planetary and surface albedo expressed as a function of latitude.

energy has maximum values of $300 \, \text{W m}^{-2}$ in low latitudes. The planetary radiation, which is dominated by emission from the lower troposphere, shows a decrease with latitude which occurs at a slower rate than the decrease in the absorbed solar radiation. If the atmosphere were in radiative balance at each latitude, the two curves should be identical. However, the solar absorption exceeds the planetary emission between 40°N and 40°S, and therefore there is a net excess in low latitudes and a net deficit in high latitudes (i.e. poleward of 40°). This latitudinal imbalance in radiation implies that heat must be transferred from low to high latitudes by the circulation of the atmosphere and ocean. In effect, the atmospheric and oceanic circulations are maintaining higher atmospheric temperatures polewards of 40° than can be maintained by solar absorption.

To obtain the transport of heat by atmosphere and ocean, the net radiation at each latitude, i.e. the difference between the absorbed solar radiation and the emitted planetary radiation, is integrated from pole to pole (figure 1.6). Thus it is possible to determine the poleward transport of heat from satellite radiation measurements, but it is not possible to distinguish what proportion of the heat is transferred by the atmosphere and what proportion by the ocean. It is noted that the maximum transfer of heat occurs between 30° and 40° of latitude and is approximately $4 \times 10^{15} \, \text{W}$.

Seasonal variations in the surface albedo are significant, particularly in the middle and high latitudes of the Northern Hemisphere, and arise

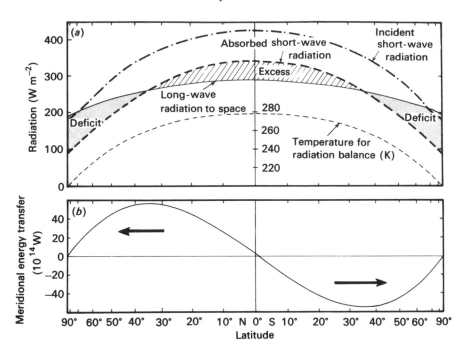

Figure 1.6. (a) Radiation balance of the Earth–atmosphere system as a function of latitude. The thin broken curve indicates an equivalent radiative temperature (centre scale) for each latitude band to be individually in balance with the absorbed short-wave radiation (heavy broken curve). The thin solid curve shows the actual observed long-wave radiation emitted to space as a function of latitude. (b) Poleward energy transfer within the Earth–atmosphere system needed to balance the resulting meridional distribution of net radiative gains (hatched) and losses (stippled) as indicated in (a). Note that the horizontal scale is compressed towards the poles so as to be proportional to the surface area of the Earth between latitude circles. (Reproduced from Climate Change and Variability: A Southern Perspective *edited by A. B. Pittock et al., p. 12, figure 2.1.2 with permission of Cambridge University Press.)*

from changes in the snow and cloud cover. At 60°N the albedo varies from 65% in February to less than 40% in July, as a result of the seasonal variation in snow cover. In the Southern Hemisphere the seasonal cycle is not so pronounced but a minimum in albedo can be observed in March at 60°S which is probably associated with the minimum sea ice extent around Antarctica. The emitted planetary radiation shows a strong asymmetry between the Southern and the Northern Hemispheres which is the result of the different land–sea distribution in each hemisphere.

The net radiation budget, illustrated in figure 1.7, indicates substantial seasonal variation as the result of the seasonal solar distribution. The zero contour varies from 15 to 70° latitude, thus showing that the zone of maximum heat transport will also vary during the seasonal cycle. The maximum net radiation gain varies between 30°S and 30°N and is larger in the Southern Hemisphere than in the Northern Hemisphere.

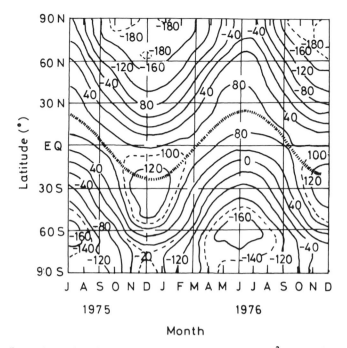

*Figure 1.7. Seasonal variation of zonal mean net radiation budget, $W \, m^{-2}$, measured by satellite from July 1975 to December 1976. (Reproduced from H. Jacobowitz et al. (1979), J.A.S., **36**, 506, figure 7 with permission of the American Meteorological Society.)*

The surface albedo, as has been shown, makes an important contribution to the planetary albedo although the cloud distribution is, without doubt, the dominant influence. However, the amount of solar radiation absorbed by the surface, and therefore the surface temperature, are strongly influenced by the albedo of that surface. The albedo of any particular surface depends on the following factors:

(i) The type of surface.
(ii) The solar elevation and the geometry of the surface relative to the Sun.
(iii) The spectral distribution of the solar radiation and the spectral reflection.

Table 1.3 shows typical values of the surface albedo. It is noted that for a calm sea surface, the albedo is only 2% and therefore the oceans are generally very good absorbers of solar radiation in comparison with most types of land surface. Forests and wet surfaces generally have low albedoes, whilst deserts and snow-covered surfaces have high albedoes. The surface albedo is also a function of the spectral reflectivity of the surface and

Table 1.3. Albedos for different terrestrial surfaces.

Type of surface	Albedo (%)
Ocean	2–10
Forest	6–18
Grass	7–25
Soil	10–20
Desert (sand)	35–45
Ice	20–70
Snow (fresh)	70–80

variations between visible and near infra-red can be substantial. For instance, though freshly fallen snow has an albedo greater than 90% in the visible, at $1 \cdot 5$ μm in the near infra-red its albedo is only 25%. Vegetation usually has a low albedo in the visible region but a high albedo in the near infra-red. Although it is useful to be able to derive a single albedo for the solar spectrum, it must be remembered that the solar spectrum incident on the surface changes with solar elevation and cloud cover and that the albedo will also change.

1.6. *The topography of the Earth's surface*

The Earth is a spheroid which is slightly distorted by rotation. This distortion amounts to approximately $0 \cdot 3$% of the Earth's mean radius, with the equatorial radius being greater than the polar radius. Because of the irregularities in the Earth's surface, geophysicists have defined a surface known as the geoid. The geoid is a surface of equipotential gravity. The height of this surface varies with position relative to the centre of mass of the Earth, but *along* the surface there is no gravitational force, i.e. the direction of the gravity force is everywhere perpendicular to the equipotential surface.

The ocean surface would be an equipotential surface if winds, tides, currents and density variations did not produce deformations in its surface. Oceanographers, therefore, have defined their reference level in terms of the mean sea level, after the removal of tides and meteorological influences. The mean sea level has been obtained by long-term observations of the sea level at coastal stations around the world. With reference to the mean sea level, the continental surface has a mean altitude of 840 m, whilst the mean depth of the ocean is 3800 m, which implies that the average height of the Earth's crust is 2440 m below sea level.

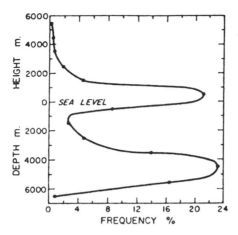

Figure 1.8. Frequency distribution of elevation intervals of the Earth's surface. (Reproduced from Elements of Physical Oceanography, *by H. J. McLellan, p. 5, figure 1.2, with permission of Pergamon Press.)*

The relative topography of the Earth's surface is depicted in figure 1.8, where it is noted that the frequency distribution has two maxima; one corresponding to the mean height of the continents and the second corresponding to an ocean depth of approximately 4500 m. This diagram shows that the most frequent height of the Earth's crust corresponds to the ocean bottom. The ocean occupies $70 \cdot 8\%$ of the Earth's surface, and 76% of this ocean area has depths between 3000 and 6000 m. These depths are known as the abyssal depths of the ocean. Of the remaining 24% of the ocean area, $7 \cdot 5\%$ is occupied by the continental shelf, which is roughly confined to depths of less than 200 m, and $15 \cdot 3\%$ is occupied by the continental slope, which is the transition region between the continental shelf and the abyssal ocean. Only $1 \cdot 1\%$ of the ocean has depths greater than 6000 m and only $0 \cdot 1\%$ has depths greater than 7000 m. This small percentage is associated with ocean trenches, whose depths plunge down to 11,500 m in the Mindanao Trench.

The relative distributions of land and ocean as a function of latitude are shown in figure 1.9. It is again noted that the land/sea distribution is asymmetrical, with over 80% of the total global land area being located in the Northern Hemisphere and 63% of the global ocean area in the Southern Hemisphere. The continental land masses of North America and Eurasia contribute to the dominance of land over ocean between 45 and $70°$N. At corresponding latitudes in the Southern Hemisphere are regions where land occupies less than 3% of the total area, with essentially no land betwen 55 and $60°$S. Poleward of $70°$ latitude the situation is reversed, with the Northern Hemisphere being occupied by

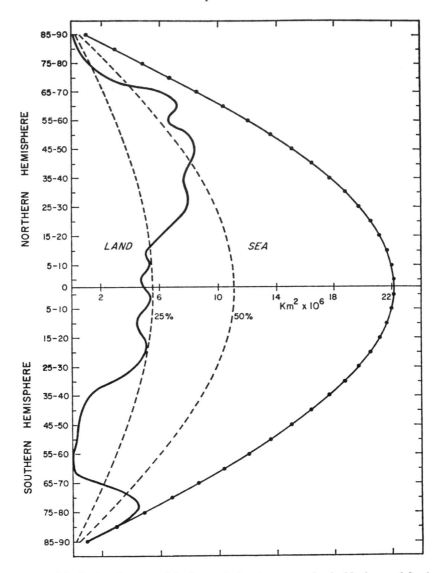

Figure 1.9. Distribution of water and land areas in five degree zones for the Northern and Southern Hemispheres. (Reproduced from Elements of Physical Oceanography, *by H. J. McLellan, p. 4, figure 1.1, with permission of Pergamon Press.)*

the Arctic Ocean and the Southern Hemisphere being occupied by the continent of Antarctica. It is believed that the presence of a continent in a polar region is a significant factor in causing glacial cycles over the past 60 million years. An interesting property of the land/ocean distribution is that 95% of all land points have antipodes in the ocean.

The distribution of the major topographic features of the globe is shown in figure 1.10. The most extensive feature is the continental slope, or escarpment, which extends for more than 350,000 km with an average height above the ocean floor of $3 \cdot 5$ km. The continental slope marks the geological boundary between the older, relatively thick, granitic continental crust and the young, basaltic ocean crust.

The second most important feature is the mid-ocean ridge system which has a length of approximately 75,000 km. The typical width of the ridge is about 500 km and the height ranges from 1 to 3 km above the abyssal depths. The ridges are regions of newly formed ocean crust, with the crust moving outwards from the central part of the ridge system. A rift valley approximately 20–50 km wide, along the central part of the system, is the region of major crustal formation and it is frequently associated with volcanic and seismic activity. The ocean ridges are of importance in the movement of deep-water masses because they split the oceans into smaller ocean basins. In the Atlantic Ocean the eastern abyssal basins are separated from the western abyssal basin by the mid-Atlantic ocean ridge. Connections between the basins, however, occur in the fracture zones which are orientated perpendicular to the axis of the ridge. Two important fracture zones in the Atlantic Ocean are the Romanche gap close to the equator, which allows Antarctic bottom water to flow into the eastern basins, and the Gibbs Fracture zone at about $53°$N, which allows the North Atlantic bottom water to flow westward into the Labrador basin. Furthermore, the ridge system has an influence on oceanic flows above the level of the ridge topography and in some regions, particularly in the Southern Ocean, its effect extends to the ocean surface currents.

The relatively low relief of the land masses has already been noted and only 11% of the land area is above 2000 m. However, there are two significant features associated with the tertiary mountain building which are worthy of further comment. These are the Western Cordillera of the American continent, which has an altitude of over 6 km above sea level, and the Himalaya mountain chain with altitudes in excess of 8 km. These mountain ranges are very important in their influence on the general circulation of the atmosphere. Although the atmosphere extends well beyond the highest peaks, 50% of the atmospheric mass occurs below 5 km and therefore these large mountain chains have a major disruptive effect on the flow of the lower atmosphere. The Western Cordillera has a predominantly north–south orientation and this disrupts the westerly flow for over 5000 km downstream. The Himalayas have an east–west orientation which results in an intensification of the zonal flow and an inhibition of the northward penetration of the Asian monsoon. Not only is the flow directly affected by the mountain barriers but because most of the water vapour in the atmosphere is confined to the lowest 5 km, the

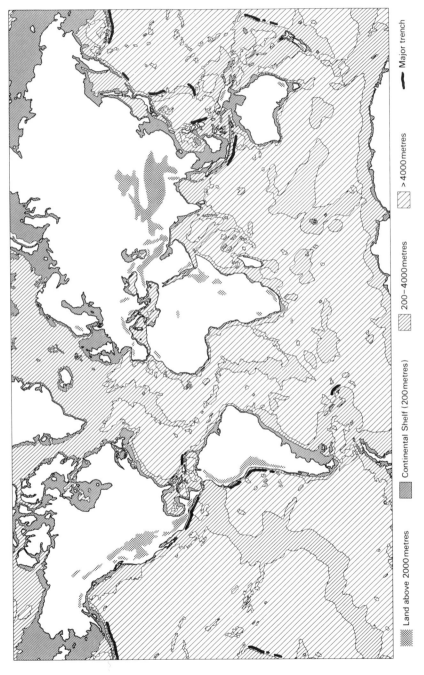

Figure 1.10. Major topographic features of the ocean and the continents.

geographical distribution of the hydrological cycle is also significantly affected. Furthermore, the release of latent heat by condensation into the atmosphere provides a major source of energy for the general circulation of the atmosphere and the mountain chains will influence the location of heat sources, in the atmosphere and the associated low-level atmospheric flow.

A feature worthy of note is the comparison of the horizontal scales of the ocean and atmosphere with the vertical scales. The ocean basins have horizontal dimensions of 5000 km in the Atlantic Ocean to over 15,000 km in the Pacific Ocean, whilst the average depths are approximately $3 \cdot 8$ km. Thus the vertical scale is small compared with the horizontal scale, and the ratio between them, known as the aspect ratio, is at most 1 part in 1000 and, more typically, 1 part in 10,000. The atmosphere has 99% of its mass between the sea surface and a height of 30 km and, compared with the circumference of the globe, the aspect ratio is again less than 1 in 1000. Thus, both atmosphere and ocean are indeed very thin when compared with their horizontal dimensions.

It has been shown that the topography of the land masses has a significant effect on the circulation of the lower atmosphere. In a different way, the shape and size of the ocean basins has profound effects on the circulation of the water masses. Except for the region lying between 50 and $60°$S, the ocean is bounded by continental masses and therefore zonal (i.e. west–east) flows are restricted to the Southern Ocean. The presence of continental margins produces intense meridional, i.e. north–south, flows in the western regions of the basins as evidenced by the Gulf Stream in the North Atlantic and the Kuroshio current in the North Pacific. All the ocean basins are connected via the circumpolar flow of the Southern Ocean. In contrast, the flow into the Arctic Ocean from the Pacific Ocean is severely inhibited by the shallowness of the Bering Strait which is less than 50 m deep, and therefore only the Atlantic Ocean is directly connected to the Arctic Ocean, between Greenland and Scotland. There is a small deep connection between the Indian Ocean and the Pacific Ocean, but this flow is, again, relatively small in comparison with the circumpolar flow in the Southern Ocean.

The width of the ocean basins has a marked effect on the ocean–atmosphere interaction. The relatively narrow North Atlantic Ocean is exposed to drier air from the surrounding continents than the Pacific Ocean and therefore evaporation is approximately twice as large in the North Atlantic Ocean as in the Pacific Ocean. This high evaporation rate results in a larger heat loss from the North Atlantic Ocean than the Pacific Ocean, and hence dense, cold water masses are formed in the North Atlantic Ocean rather than in the Pacific Ocean. Thus the deep-water vertical circulation in the North Atlantic Ocean is much more intense than the sluggish vertical circulation observed in the Pacific Ocean. It will be shown in Chapter 5 that this has consequences for the heat transport in the ocean. In the context of evaporation, the marginal seas, such as the Norwegian Sea, the Mediterranean Sea and the Red Sea, are important regions for the formation of dense water masses because of their high evaporation rates.

The continental shelf, though occupying a relatively small area of the ocean, is of considerable importance both to man and to life in the ocean. The intense flows, associated with winds and tides in the continental shelf region, result in a mixing of the nutrients from the continental shelf sediment into the water column. These nutrients, in conjunction with the ready availability of oxygen and sunlight in these shallow sea areas, produce an abundance of phytoplankton. These regions of high primary production are very important in the maintenance of marine food chains and contain the most intensive fisheries in the world. The sedimentation rates on the continental shelves are also high, being of the order of 1 m/1000 years in comparison with 0·1–10 mm/1000 years in the deep ocean. Much of the sediment deposited is derived from river-borne material and this contains large quantities of phosphorus, which is an element necessary in the maintenance of productivity in the ocean.

In conclusion, it has been shown that the surface topography and the shape and interlinkage of the ocean basins have a significant effect on the circulation of both the atmosphere and the ocean, which, in turn, affect the biological productivity of the land and the ocean. In the geological past, different ocean geometry and surface topography would have been partly responsible for the many varied climatic patterns which have occurred since the formation of the Earth.

2

Composition and physical properties of the ocean and atmosphere

2.1. Evolution of the atmosphere and ocean

It is thought that the Earth evolved from the accretion of planetismals whose gravity fields were too weak to retain a primordial atmosphere. The evidence for this theory comes from the study of the composition of the Earth's atmosphere and the comparison of this composition with other planetary atmospheres and cosmic abundancies. It is found that the rare gases are present in the Earth's atmosphere in abundancies which range from only 10^{-10} to 10^{-6} of their observed cosmic abundancies. There is a relative abundance of the lighter elements of hydrogen and helium on the Sun and the massive planets and a paucity on the inner planets. This shows that the inner planets lost the greater proportion of the mass attributable to the normal cosmic abundancies of the gaseous elements. This evidence suggests that the Earth's atmosphere evolved from entirely self-contained, secondary sources.

Volcanic activity is probably the most important source of the early atmosphere. Gaseous emission of H_2O, CO_2, N_2, SO_2, SO_3, H_2 and Cl_2 occurred in substantial quantities, though it is important to note that no oxygen was, or is, directly released from volcanic effluent. Observations from present-day volcanoes indicate that H_2O, CO_2 and SO_2 are the primary effluents. At the same time, solid effluent built up the continents at a rate of $1-3$ km^2 per year.

The evolution of this carbon-dioxide-dominated atmosphere, which had characteristics very similar to the atmospheres now present on Venus and Mars, to an oxygen–nitrogen atmosphere is the central atmospheric evolution problem. Of particular importance is the question of the origin of oxygen in the present-day atmosphere and the disappearance of carbon dioxide. Evidence for the lack of oxygen in the early atmosphere is based on:

(i) The lack of geological sources of oxygen gas.
(ii) The incomplete oxidation of early sedimentary materials laid down about 3×10^9 years ago.

In the primitive atmosphere, the oxygen concentration was probably controlled by the photodissociation of water molecules by ultraviolet radiation, which would penetrate to the surface in the absence of a significant ozone layer, thus:

$$H_2O + h\nu \rightarrow H + OH$$
$$H + OH + h\nu \rightarrow 2H_2 + O$$
$$O + O \rightarrow O_2$$

This gave rise to an oxygen level of 10^{-4} of the present atmospheric level (PAL). The reason for the low concentration of oxygen gas is that it would tend to mix upwards through the thin atmosphere, whilst the water vapour would easily condense out at all but the very lowest levels of the atmosphere. The oxygen would have tended to shield the water molecules from further ultraviolet radiation and therefore inhibit photodissociation. In such an atmosphere, ozone (O_3) would also be produced but not in sufficient quantity to absorb all the incident ultraviolet radiation. This ozone would probably have been removed by surface oxidation by a similar process to that observed on Mars.

The subsequent rise in oxygen concentration can be attributed only to photosynthesis, i.e. the oxidation of liquid water and the reduction of carbon dioxide to form carbohydrate:

$$6H_2O + 6CO_2 + h\nu \rightarrow (C_6H_{12}O_6) + 6O_2$$

The oxygen produced by photosynthesis and by the photodissociation of water is also removed by oxidation at the surface, decay and respiration, and by dissolution in water. Only when more oxygen is produced than is being removed can the development of an oxygen atmosphere take place.

The process of photosynthesis requires the protection of simple cells from the lethal ultraviolet radiation which was able to penetrate to the surface of the Earth through the primitive atmosphere. Such protection may be obtained from a depth of a few metres of liquid water, which is sufficient to absorb the damaging ultraviolet radiation. However, visible light is also absorbed by liquid water and at a depth of 30 m most of the visible light has been absorbed. Thus the primitive cells would have been able to photosynthesize only in a relatively thin layer. The energy for these primitive cells would probably have come from organic nutrients synthesized at the surface by the incident ultraviolet radiation which were then mixed down. The growth of the oxygen gas concentration would have been slow prior to the attainment of a certain critical value, at which time the penetration of the lethal ultraviolet radiation, now incident on the Earth's surface at a much-reduced intensity due to the development of the ozone layer, would have diminished to a few centimetres of water. Secondly, the Pasteur point would have been reached when organisms change from fermentation to respiration. This would occur at $0 \cdot 01$ PAL which coincides with the Cambrian Era some 600×10^6 years ago. Before this time, there is evidence only for algae, fungi and bacteria and no fossil records. At this stage, the development of an oxygen atmosphere would have proceeded at an explosive rate as marine species multiplied rapidly.

Once oxygen concentrations had reached $0 \cdot 1$ PAL, there would have been sufficient ozone in the atmosphere to shield the land from the harmful ultraviolet radiation and therefore life would have been able to spread onto land. This occurred some 400×10^6 years ago. The rapid growth of plants and animals in a period of 30×10^6 years resulted in a 25% increase in oxygen levels. The high rates of photosynthesis, coupled with a lack of decay products on land, not only led to a substantial increase in oxygen levels but also to a corresponding decrease in carbon dioxide concentration. Thus oxygen gas concentrations reached a maximum in the Carboniferous period. At this stage, a substantial decrease in the global surface temperature occurred and the ice ages of the Permian period developed some 270 million years ago. This cooling resulted in a net decrease in photosynthesis and a subsequent decrease in the oxygen concentration to close to present levels.

The carbon dioxide in the primitive atmosphere was removed successively by the production of organic material, which was laid down in sediments, and by the formation of carbonates in the early ocean. The inventory of the present carbon budget (table 2.1) shows that by far the largest quantity of carbon is present in the form of carbonate rocks and shales. The ocean contains a relatively large amount of carbon as the result of dissolution of carbon dioxide gas, whilst the biosphere and

Table 2.1. The carbon budget for the Earth.

Reservoir	Mass of carbon (10^{14} kg)
Atmosphere	7
Ocean	380
Land	
Vegetation	8
Soil	15
Sediments	200 000
Fossil fuels	50

atmosphere contain rather small reservoirs of carbon. It is interesting to note that the burning of a large proportion of the fossil fuels could lead to an increase in atmospheric carbon dioxide many times that of the present levels. It is unlikely that the biosphere could take up this excess of carbon dioxide, although the presence of a substantial potential reservoir for carbonates in solution in the ocean might make possible the accommodation of much of the excess carbon dioxide. However, because of the lack of knowledge about the rates of uptake of anthropogenic (i.e. man-made) carbon dioxide by the ocean and by photosynthesis it is not known what levels of carbon dioxide would be attained by the atmosphere. A fourfold increase in present carbon dioxide concentration in the atmosphere could lead to a global increase in temperature of approximately 4 K, and cause a major shift in the world's climatic zones.

The development of the ocean

The evolution of the ocean is ultimately connected with the evolution of the atmosphere and the continents. The ocean is probably unique in our solar system, although Saturn's moon, Titan, may have a deep ocean consisting of liquid methane and acetylene, and is attributable to the Earth being at such a distance from the Sun that its surface temperature and pressure allow water to exist in liquid form. The high temperatures and pressures of the Venusian atmosphere allow only the gaseous form of water to exist, whilst on Mars the low temperature and pressure preclude the presence of large quantities of liquid water, although there is evidence from the geological studies of Mars that liquid water was present earlier in its history. The Earth's atmosphere can only hold a relatively small amount of water as vapour because of the rapid decrease of temperature and pressure with height. Therefore the majority of the water vapour ejected by volcanic activity would have been quickly transformed into the liquid phase in the early phase of Earth's evolution. It has been speculated that the source of water was the crystallization of the magma into igneous rock and this process could account for the total water content of the ocean. Escape may have been from hot springs which are more numerous than volcanoes. Although most of the water emitted from hot springs and volcanoes is recycled ground water, which has seeped down from the surface, it can be shown that it would only require 1% of this water to be 'juvenile' water, at present emission rates, to produce the present-day volume of the ocean in 4×10^9 years. If this theory is correct, the ocean would have been steadily increasing in volume since the formation of the Earth. However, there is no evidence for the flooding of the continents in the geological record and therefore it is believed that the continents increased in volume at about the same rate that new water was added. Because of isostatic control, the depth of the ocean, the sea level and the continents must have remained in equilibrium throughout geological time.

The present-day ocean basins are all comparatively young with respect to the age of the Earth ($4 \cdot 6 \times 10^9$ years old). There is no evidence from these basins of sediments older than 190×10^6 years, and over half these sediments are younger than 75×10^6 years. More ancient sediments are only found on the continents and continental margins, the remainder having been removed into the Earth's mantle along regions of descending motion, known as subduction zones. Thus the ocean crust is much younger than the continental crust. New ocean crust, composed of basaltic igneous rocks, is formed along spreading centres located along the mid-ocean ridges. The rate of spreading varies between 1 and 8 cm per year and is fastest at the East Pacific Rise and slowest at the Mid-Atlantic Ridge. At the present rate of production of oceanic crust along the East Pacific Rise, the Pacific Ocean basin could have been formed in approximately 200×10^6 years. The oldest sea floor so far determined in the Pacific Ocean is indeed about 150×10^6 years old. Estimates for the Atlantic Ocean indicate a similar age to the Pacific Ocean. The Red Sea and the Gulf of Aden is thought to be an embryonic ocean basin which started spreading about 20×10^6 years ago.

2.2. *Present-day composition of sea water*

Dissolved salts comprise about $3 \cdot 5 \%$ by volume of sea water. The salts are present because of the relative ease with which ionic compounds dissolve in water. Originally it was believed that the salts dissolved in the ocean were derived from the weathering of continental rocks, the products of which were then transported by rivers to the ocean. Robert Boyle made measurements of the composition of river water and he showed that salts could be detected. However, with more accurate methods of analysis, it was found that although the river waters contained potassium, magnesium, calcium and sodium ions they were deficient in dissolved carbon dioxide (i.e. carbonate and bicarbonate), chlorides, bromides and iodides. In fact, if the ocean salts had been derived from this one source, their composition would have to be totally different from that observed (table 2.2). The ions missing from land weathering products originated from volcanic and hot spring discharges in a similar manner to the way water was emitted at a steady rate over geological time.

The observed distribution is not, however, simply determined by the inputs from weathered rock and volcanic emissions, because ions are simultaneously being lost from the water column by sedimentation on the ocean floor and continental shelves and by transfer to the atmosphere and

Table 2.2. Composition of standard ocean water (salinity = $34 \cdot 7‰$).

Constituent ion	Symbol	Mass of ion in grams for 1 kg of sea water	Percentage of total salt
Chloride	Cl^-	$19 \cdot 215$	$54 \cdot 96$
Sodium	Na^+	$10 \cdot 685$	$30 \cdot 58$
Sulphate	SO_4^{2-}	$2 \cdot 693$	$7 \cdot 70$
Magnesium	Mg^{2+}	$1 \cdot 287$	$3 \cdot 69$
Calcium	Ca^{2+}	$0 \cdot 410$	$1 \cdot 17$
Potassium	K^+	$0 \cdot 396$	$1 \cdot 13$
Bicarbonate	HCO_3^-	$0 \cdot 142$	$0 \cdot 41$
Bromide	Br^-	$0 \cdot 067$	$0 \cdot 19$
Boric acid	H_3BO_3	$0 \cdot 026$	$0 \cdot 07$

biosphere. The present-day composition is, therefore, a geochemical balance between the sources and sinks of the ions. It is believed that the ocean has been in equilibrium for 100×10^6 years and that the total salt content of the ocean has not changed significantly in that time. This is based on the fact that the pH of the ocean must have remained reasonably constant to enable the present life forms to evolve. It is noted that a 1% decrease in the sodium ion concentration would result in a strongly acidic ocean with a pH of $2 \cdot 6$. However, the ratios of the constituent salts in the ocean have shown considerable fluctuations. For example, the Na^+/K^+ ratio was approximately 1 in the pre-Cambrian period ($\sim 10^9$ years ago) and now it is 28. The changes in the chemical composition are related to the evolution of organisms which extracted materials such as calcium and silicon for shell building and so altered the steady-state balance.

The first comprehensive analysis of the composition of sea water was made by William Dittmar who analysed 77 samples of sea water, collected by H.M.S. *Challenger* between 1872 and 1877. The samples were representative of all the major ocean basins and were taken at depths ranging from the surface to the abyssal layers. Dittmar analysed the samples for the eight major constituents (see table 2.2) and showed that, within narrow limits, the ratios of these major constituents were in constant proportion. Together with boric acid, these eight constituents comprise $99 \cdot 9\%$ of the total dissolved salts in the ocean. The remaining $0 \cdot 1\%$ represents virtually all the stable elements of the periodic table and they do not generally obey the rule of constant proportions. The most prominent of the minor constituents are strontium, silicon, iron, lithium, phosphorous and iodine, as well as the dissolved gases oxygen and nitrogen, all of which are fundamental to the chemical and biological processes in the ocean.

Dittmar's experimental proof that the major constituents were present in constant proportions to each other throughout the world's oceans, including the Mediterranean and Baltic Seas, allowed oceanographers to define the salt content of the ocean as a single physical quantity, known as salinity. From a determination of the salinity and the rule of constant proportions, the concentration of any one of the major constituents can be determined. Furthermore, in order to obtain a value of the salinity of sea water, it is only necessary to measure one of the major constituents. The most easily measured constituent is the chloride content of sea water. In practice, the chlorinity (Cl) is determined, which is the amount of chlorine (grams) in 1 kg of sea water, where both bromine and iodine are replaced by chlorine. The relationship between salinity and chlorinity is given by

$$\text{salinity } (\text{\textperthousand}) = 1 \cdot 80655 \text{ Cl}$$

By titration against silver nitrate, chlorinity can be measured and the salinity obtained to an accuracy of $\mp 0 \cdot 01$.

Recently this technique has been replaced by direct physical techniques which measure the electrical conductivity or the refractive index of the ionic solutions. These techniques are both more accurate and more sensitive ($\mp 0 \cdot 003$) than the titration method.

2.3. Introduction to gases and liquids

The relationship between pressure, p, temperature, T, and the density, ϱ, for an ideal gas is:

$$p = \varrho R T$$

where R is the gas constant for the constituent gas having molecular weight m. $R = R^*/m$, where R^* is the universal gas constant.

The derivation of the ideal gas equation depends on the assumption that the molecules in an ideal gas are so far apart that intermolecular forces can be disregarded except during the very brief collisions. The strong intermolecular forces between ideal gas molecules allow them to be treated as hard spheres and their collisions may be considered to be inelastic. Many gases approximate to ideal gases under certain temperature and pressure conditions.

Such simple assumptions are invalid when dealing with the liquid and the solid state. In a liquid the intermolecular forces cannot be neglected nor generalized, since they vary from one liquid to another. It is not, therefore, possible to produce an ideal equation of state for an ideal liquid. The long-range attractive intermolecular forces in a liquid allow a much more efficient packing of the molecules than is found in a gas. Hence, liquids have densities approximately 100–1000 times greater than the density of a gas. For instance, air at normal atmospheric pressure and $0°C$ has a density of $1·28 \, kg \, m^{-3}$ whilst sea water has a density of $1028 \, kg \, m^{-3}$.

At very short ranges, of the order of $0·1 \, nm$, the intermolecular repulsive forces become dominant, and there is, therefore, a limit to the close packing of molecules. Solids may have densities approximately 15% greater than liquids but this increase is achieved by more efficient, long-range packing arrangements rather than a smaller intermolecular separation. In this respect, the densities of water and ice are anomalous. Ice has a density which is 9% less than liquid water and thus ice can float on water.

The liquid state has properties similar to the solid state on a short time scale in that small clusters of molecules are ordered in a similar way to the solid-state lattice. However, on a large time scale the liquid resembles a gas in that it is highly disordered. Liquids may therefore be considered, in some respects, as being an intermediate state between the extreme ordering of the solid-state lattice and the randomness of the gaseous state. On this longer time scale a liquid and gas will not support a shear stress, and as a result they will yield and flow. In this case we may call the gas or liquid a fluid and we can then describe mathematically properties of motion without referring to the substance as liquid or gas.

Gases, furthermore, have the property that they will expand to fill the space available, and therefore it is impossible for a gas to have a free surface. In this context it is not meaningful to refer to the height or thickness of the atmosphere. A liquid on the other hand always has a free

Table 2.3. Composition and scale heights of terrestrial atmospheres.

	Major constituents (% of molecules)	Mean molecular weight (a.m.u.)	Gas constant $(J \, kg^{-1} \, K^{-1})$	Scale height (km)
Venus	95·5% CO_2 4% N_2	43·35	192	14·9[a]
Earth	78% N_2 21% O_2	28·97	287	7·4
Mars	96% CO_2 2·5% N_2 1·5% Ar	43·55	191	10·9

[a]Lower atmosphere.

surface, and therefore it is possible to define the interface between atmosphere and ocean. Another important property of a gas is that it is readily compressible due to the large distances between molecules. However, because of the close packing in a liquid it is difficult to compress a liquid, without resorting to large pressures.

The atmosphere can be described as a mixture of gases, predominantly oxygen, nitrogen, carbon dioxide, argon and water vapour (table 2.3). Of these major constituents only water vapour is a significantly variable constituent and it is therefore considered separately to the other gases. The equation of state for dry air can be obtained by summing the partial pressures of each constituent gas. If each gas occupies a similar volume, V, and has a mass of M_n then the partial pressure P_n is given by

$$P_n = \left(\frac{R^*}{m_n}\right) \frac{M_n}{V} T$$

where m_n is the molecular weight of the gas.

The total pressure of the gases is

$$p = \sum_{n=1}^{N} P_n = \frac{R^* T}{V} \sum_{n=1}^{N} \frac{M_n}{m_n}$$

The mean density of the mixture, $\bar{\varrho} = \sum_{n-1}^{N} (M_n/V)$. Therefore $p = \bar{\varrho}(R^*/\bar{m}) T$, where

$$\frac{1}{\bar{m}} = \frac{\displaystyle\sum_{n=1}^{N} (M_n/m_n)}{\displaystyle\sum_{n=1}^{N} M_n} \tag{2.2}$$

or $p = \varrho R_d T$, where R_d is the gas constant for dry air. R_d has a value of $287 \, \mathrm{J \, K^{-1} \, kg^{-1}}$.

Comparing equation 2.2 with 2.1, it is noted that the equation of state for a mixture of gases is similar to a single constituent equation of state, provided that the molecular weight is replaced by the mean molecular weight for the mixture.

Example for the Earth

Constituent	M_i	m_i
N_2	75·51	28·016
O_2	23·14	32·00
Ar	1·28	39·24
CO_2	0·02	44·01

For the constituents

$$1/\bar{m} = \frac{75 \cdot 51/28 \cdot 016 + 23 \cdot 14/32 \cdot 00 + 1 \cdot 28/39 \cdot 24 + 0 \cdot 02/44 \cdot 01}{99 \cdot 95}$$

hence $\bar{m} = 28 \cdot 96$

The equation of state for liquid water is a function of three parameters, temperature, pressure and density. For sea water, the relationship involves the properties of all the constituent salts and therefore is
extremely complicated. However, the observation of the constancy of
composition for the major elements of sea water allows the individual
constituents to be replaced by one variable, the salinity. As discussed
earlier there is no general statistical theory for all liquids and therefore the
relationship has been determined by laboratory measurements and the
results are tabulated. The *in situ* density of sea water can then be given
$\varrho = \varrho(T, P, S)$. Table 2.4 shows the values of *in situ* density for extreme
conditions in the ocean. It is noted that *in situ* density increases both with
increasing pressure and salinity, but decreases with temperature.
However, the variations in density are less than 7%, and therefore it is
common practice to define

$$\sigma(T, P, S) = \varrho(T, P, S) - 1000 \text{ kg m}^{-3}$$

The values of $\sigma(T, P, S)$ can be obtained from standard tables (e.g.
Knudsen) or from polynomial relationships. However, for many
purposes the effect of pressure is eliminated by using a reference pressure,
i.e. mean atmospheric pressure, to define density. This quantity is
referred to as sigma–t or σ_t.

Figure 2.1 shows the relationship of σ_t as a function of temperature
and salinity at atmospheric pressure. Though in all cases density increases with increasing salinity and decreasing temperature, the relationship is non-linear. At temperatures of $25-30\,^{\circ}$C typical of the equatorial
ocean the isopycnals (lines of constant density) are reasonably linear, and
a simple relationship for σ_t can be used. However, at low temperatures

Table 2.4. Density of ocean sea water for extreme ranges of temperature
and pressure in the ocean.

Depth (m)	Pressure (Pa)	Density (kg m^{-3})	
		$T = 273$ K, $S = 35‰$	$T = 303$ K, $S = 35‰$
0	10^5	1028	1022
10,000	10^8	1071	1060

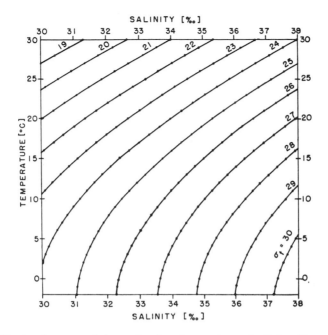

Figure 2.1. Density of sea water (σ_t) as a function of temperature and salinity. (Reproduced from H. Friedrich and S. Levitus (1972), J.P.O., 2, 516, figure 4, with permission of the American Meteorological Society.)

a given salinity change produces a larger change of density than the corresponding change in salinity at high temperature. This physical property of the equation of state results in a greater sensitivity of density to salinity changes at high latitudes, and is an important influence in the formation of bottom water. Another interesting property of the equation of state is that two water masses with the same density, but having a different temperature and salinity (see figure 5.14), when mixed will produce a denser water mass than the original water masses. Again this is an important process in the formation of dense bottom waters in polar regions of the ocean. It is amusing to speculate that if the ocean was well mixed in temperature and salinity, then the non-linearity of density would result in the increased density of the world's oceans, and an inevitable decrease in sea level.

Figure 2.2 shows T, S and σ_t for a meridional section of the Atlantic Ocean, where it is noted that the temperature effect on density tends to dominate over that of salinity at most latitudes. Only in polar latitudes is the salinity influence on density greater than that of temperature. Here fresher surface water derived from run-off from the surrounding continental land masses overlies the warmer, more saline ocean water. It can also be seen that the higher density surfaces of the

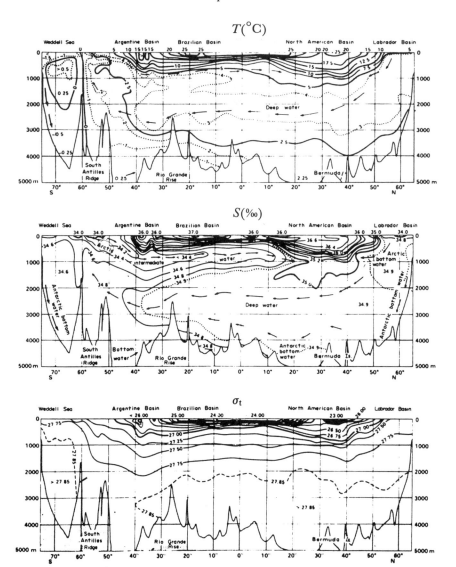

Figure 2.2. Meridional sections of T, S *and* σ_t *in the western basins of the Atlantic. (Reproduced from* Descriptive Regional Oceanography, *by P. Tchernia, Pergamon Marine Series Vol. 3, p. 162, figure 5.40, with permission of Pergamon Press.)*

deep water masses (e.g. $\sigma_t = 27 \cdot 50$) outcrop only in limited regions of the North Atlantic and in the Antarctic, where the deep-water masses are formed.

The anomalous behaviour of pure water compared with other liquids has already been briefly mentioned. In particular the maximum density of liquid water at $4\,^{\circ}$C and the expansion of water on freezing are two of its remarkable properties. Other properties include the high dielectric constant, which results from the high electrical dipole with a relatively small molecular volume. Because the interionic force of attraction varies inversely with the dielectric constant, the attraction between ions is diminished in water, and hence ionic components are easily dissolved in water. The presence of the dissolved salts in sea water increases the electrical conductivity by 10^6 over that of pure water. This property has been used to advantage in the accurate electrical measurement of salinity in sea water, which is now the most reliable technique available to oceanographers. Additional remarkable features of water are its high specific heat, its high boiling point compared with other hydrogen-based liquids (e.g. 213 K for H_2S and 240 K for NH_3) and its large latent heat of vaporization of $2 \cdot 3 \times 10^6 \, J \, kg^{-1}$, compared with $0 \cdot 55 \times 10^6 \, J \, kg^{-1}$ for H_2S and $1 \cdot 4 \times 10^6 \, J \, kg^{-1}$ for NH_3. This property allows water to be a major factor in the transfer of energy between the ocean and atmosphere.

The influence of salinity on the properties of water is however significant. Figure 2.3 shows the temperature of the maximum density of sea water and the freezing point as a function of salinity. For salinities less than 24·69‰, water reaches a maximum density before the freezing point is reached, and therefore between the temperature of maximum

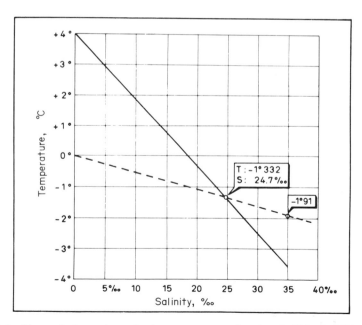

Figure 2.3. Changes in the maximum density temperature as a function of salinity (solid line). Changes in the freezing point as a function of salinity (dotted line). (Reproduced from Descriptive Regional Oceanography, *by P. Tchernia, Pergamon Marine Series Vol. 3, p. 35, figure 3.4, with permission of Pergamon Press.)*

density and the freezing point there is an anomalous decrease in density with decreasing temperature. However, for a water mass having a salinity in excess of 24·69‰, which is true of all ocean water, the freezing point is reached before the maximum density of sea water. In this case, there is no anomalous behaviour of sea water, and the densest water masses (for a given salinity) are associated with the lowest surface temperatures. For most ocean water masses this temperature $\sim -2°C$. A second property of salinity is the effect on the specific heat of water. For pure water, at $0°C$, the specific heat is $4218 \, \mathrm{J \, kg^{-1} \, K^{-1}}$, whilst for sea water it is $3940 \, \mathrm{J \, kg^{-1} \, K^{-1}}$ for a salinity of 35‰, and therefore sea water is 6% less efficient in storing heat than an equivalent mass of fresh water.

The propagation of sound in sea water is given by $C = \sqrt{K/\varrho}$, where ϱ is the density and K is the compressibility of sea water. They are both dependent on the temperature, salinity and pressure. Figure 2.4 shows the variation in sound velocity for typical temperature and salinity profiles in the ocean. In the deep ocean (> 1500 m), because of the uniformity of temperature and salinity, the sound velocity is dominated

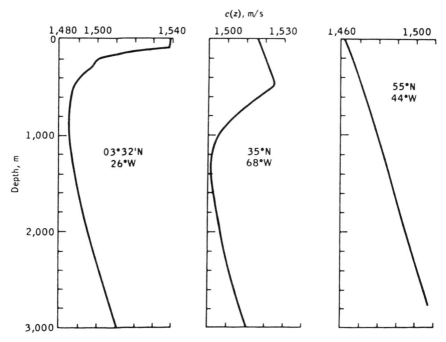

Figure 2.4. Velocity of sound as a function of depth, for equatorial, sub-tropical and sub-polar oceans. Between the surface and 1000 m, temperature is the dominant effect, while pressure becomes dominant at depths greater than 1000 m. (Reproduced from Ocean Acoustics: Theory and Experiment in Underwater Sound, *by I. Tolstoy and C.S. Clay, p. 7, figure 1.2, with permission of McGraw-Hill.)*

by pressure and therefore increases with depth. In the upper ocean, where large variations of temperature, and to a lesser extent salinity, occur, the sound velocity displays more variability. The large decrease of temperature with depth, especially through the thermocline, causes a decrease in velocity with depth which is larger than the opposing pressure effect. Hence a minimum velocity is observed at a depth of approximately 1 km in profiles where the thermocline is present. This minimum is very important for the propagation of sound and is known as the SOFAR channel. Sound waves emitted either upwards or downwards from this level will exhibit refraction into the sound channel, because of the variation of the velocity of sound with depth. In this region, therefore, sound energy is trapped in a wave guide, and a signal may propagate over hundreds and even a thousand kilometres before being significantly attenuated. This property has been used by oceanographers for the distant monitoring of Swallow floats and SOFAR floats, from research ships and shore bases.

2.4. Hydrostatic equilibrium

An observer descending from the outer reaches of the atmosphere to the Earth's surface with a barometer would note an increase in pressure. This increase in pressure would not be linear but would show an exponential increase with depth. Though pressure itself is a scalar quantity, the vertical variation of pressure in the atmosphere produces a pressure force which acts in the direction of decreasing pressure (i.e. upwards). This observed pressure-gradient force would be able to accelerate a parcel of air upwards at ~ 10 m s^{-2}. In the absence of other forces, this acceleration would produce a velocity of ~ 5000 km hour^{-1} at a height of 100 km. Though the atmosphere is always in a state of motion, measured vertical velocities are rarely larger than 10 m s^{-1}, and therefore this upward pressure gradient must be close to balance with the gravitational force acting towards the centre of the Earth. This balance between the vertical pressure gradient and the gravitational force is known as hydrostatic equilibrium. Ideally the state requires no motion and therefore the atmosphere is never quite in hydrostatic balance. However, the approximation is very good for large scales of motion and only breaks down in regions of large vertical acceleration associated with thunderstorm updrafts and mountain waves.

Consider a slab of atmosphere of thickness δz and cross-section area δA. The pressure force across the slab is $[p - (\partial p/\partial z)\, \delta z]\, \delta A - p\, \delta A$

$$\text{PF} = -\frac{\partial p}{\partial z} \delta z\, \delta A$$

The force due to gravity on the element of atmosphere is $\varrho \, \delta A \, \delta z g$ hence the balance is $(\partial p / \partial z) \, \delta z \, \delta A = - \varrho \, \delta A \, \delta z g$ or

$$\frac{\partial p}{\partial z} = - \varrho g \qquad (2.3)$$

The hydrostatic equation 2.3 is a very important relationship in both meteorology and oceanography because it relates the pressure to the depth of fluid. In the atmosphere, however, the density also varies with height, because of the compressibility of gases, and therefore equation 2.3 is not sufficient. However, since the atmosphere is a very good approximation to an ideal gas, the ideal gas equation can be used to eliminate density in favour of temperature and pressure. On substitution of $\varrho = p / RT$, we have

$$\frac{\partial p}{\partial z} = - \frac{pg}{RT}$$

If T is replaced by the vertical mean temperature of the atmosphere $\bar{T}$, the equation can be integrated from the surface ($z = 0$), to yield the pressure at a height z',

$$\int_{p_0}^{p'} \frac{\partial p}{p} = - \frac{g}{R\bar{T}} \int_0^{z'} \partial z$$

or

$$p' = p_0 \exp(- z' / H) \qquad (2.4)$$

where p_0 is the surface pressure, and $H = R\bar{T}/g = R^*\bar{T}/g\bar{m}$. This equation is sometimes known as the barometer equation. The quantity H is known as the scale height and is a measure of the depth of the atmosphere. When $z' = H$ then $p' = p_0 e^{-1}$ or $p' = 0 \cdot 37 p_0$. Therefore H is the height at which the pressure is 37 % of the surface atmospheric pressure. Furthermore the pressure decreases by $1/e$ for each increase in height equal to the scale height.

Table 2.3 gives typical scale heights for Earth's atmosphere and other planetary atmospheres, assuming the mean temperature is equal to the radiative equilibrium temperature. It is noted that the scale height is related to mean molecular weight of the atmosphere and the acceleration due to gravity. On Mars, for instance, the gravity is approximately one-third of the Earth's gravity, but the atmosphere is composed of carbon dioxide with a molecular weight of 44, and hence there is only a small increase in scale height. On Venus, the large vertical variation in temperature makes the scale height a highly variable parameter. The scale height is largest in the lowest layers of the Venusian atmosphere (table 2.3), where the temperature approaches 760 K (table 1.1). On Jupiter, despite the low radiative temperature, the scale height is large, because of the predominance of hydrogen and helium gas, with molecular weights of 2 and 4, respectively.

As the reader will appreciate the scale height is not a constant for a given atmospheric composition because of variations in the temperature of the atmosphere. To obtain accurate results from the equation it is necessary to measure the temperature profile of the atmosphere as a function of height, and then to integrate the equation numerically. A simpler alternative to this procedure is to use a standard atmospheric temperature profile to obtain the pressure–height relationship. This was used to calibrate pressure altimeters for aircraft, but these have been superceded by more accurate radar altimeters.

In the ocean, the relationship between pressure and depth is very close to linear, because of the small variation of density with depth. In the surface layers of the ocean, to a depth of 1 km, it is normal practice for the compressibility of sea water to be ignored. However, because of the exceptionally high pressures (100–1000 atm pressure) existing in the deep water, compressibility does become an important effect when considering deep-water masses. An illustration of the importance of compressibility is that sea level would be 30 m higher than it is today, if the ocean was assumed to be incompressible.

The pressure at a depth h in the ocean is obtained by integration of equation 2.3 from the surface to the depth h. Strictly speaking, this is not as easy as you might think because the surface of the ocean varies in height as the result of tides and large-scale water movements. The variations in sea level may be as large as 1 m. Second, the surface atmospheric pressure can vary by as much as 10% of the mean surface pressure, which will produce horizontal pressure variations in the ocean. However, for the calculation of total pressure in the ocean, these effects are relatively small, and often may be neglected although in Chapter 7 it will be shown that these small changes in pressure are important in determining horizontal currents. As a general rule of thumb, 10 m of sea water is equivalent to a pressure of 1 atm or 10^5 Pa.

2.5. Adiabatic changes and potential temperature

Consider a balloon which is thermally insulated, so that no heat is lost through the surface of the balloon, and the balloon is filled with hydrogen and released into the atmosphere. At the moment of release it is assumed that the hydrogen has the same temperature as the surrounding air. As the balloon ascends, it will expand because of the reduction of atmospheric pressure with height, and at the same time the hydrogen in the balloon will become cooler. This reduction in temperature is solely due to reduction in the internal energy of the gas, as a result of energy used to expand the surface of the balloon. Since no heat is gained or lost to the hydrogen gas, by thermal conduction across the surface or by thermal radiation from the atmosphere or Sun, the balloon is known as an *adiabatic* system, and the reduction in temperature is known as adiabatic cooling. Of course, if there is no transfer of energy to its surroundings, and the balloon is brought back down to the surface, at the same initial pressure, then the temperature of the hydrogen gas will be the same as the initial temperature. Therefore a property of an adiabatic system is that the system is reversible, and therefore no matter which path the balloon takes through the atmosphere when it returns to its initial pressure level, it will have the same initial temperature.

These properties of the adiabatic system can be understood by the application of the first law of thermodynamics:

$$\underset{\substack{\text{heat added} \\ \text{to the system}}}{dQ} = \underset{\substack{\text{internal} \\ \text{energy}}}{C_v \, dT} + \underset{\substack{\text{work done} \\ \text{by expansion}}}{p \, d\alpha} \qquad (2.5)$$

The first term represents the heat added across the boundaries of the system, which could be associated with heat conduction across the balloon's surface, and radiational heating. The second term is the internal energy of the gas, which is associated with the average kinetic energy of the molecules, and proportional to the temperature of the gas. C_v is the specific heat of the gas at constant volume. The third term is the work done on the surroundings, and in our example is associated with expansion of the volume of gas in the balloon. Hence heat added to the system may either cause a change in volume of the gas, or it may result in a change in temperature of the gas, or it may be a combination of both effects. As mentioned earlier, for an adiabatic system there is no heat added across the boundaries of the sytem and therefore $dQ = 0$. Hence it can be seen from equation 2.5, that an expansion of the balloon ($d\alpha > 0$) will cause a reduction in the internal energy of the gas and a decrease in temperature.

Before this relationship can be usefully used it is necessary to substitute for the change in specific volume, $d\alpha$. By differentiating the ideal gas equation by parts it can be shown that $\alpha \, dp + p \, d\alpha = R \, dT$. By substituting for $p \, d\alpha$ in equation 2.5 and noting $C_p = C_v + R$, where C_p is specific heat at constant pressure, then $0 = C_p \, dT - \alpha \, dp$ hence:

$$\frac{dT}{dp} = \frac{\alpha}{C_p} = \frac{1}{\varrho C_p}$$

since ϱ depends on pressure this is not a particularly useful relationship.

However, by rewriting the equation as

$$\frac{dT}{dz}\frac{dz}{dp} = \frac{1}{\varrho C_p}$$

and substituting for the hydrostatic equation we have

$$\frac{dT}{dz} = -\frac{g}{C_p} \qquad (2.6)$$

The variation of temperature with height in an adiabatic system is therefore determined solely from the value of g and the specific heat of the gas at constant pressure. For the lower atmosphere (< 10 km), the change in g with height is small and therefore the adiabatic temperature gradient is, to a good approximation, a constant. Table 2.5 shows calculated adiabatic temperature gradients for the Earth and other planets. It is noted that both Earth and Venus have similar adiabatic temperature gradients, despite the different composition of their atmosphere and masses. It is necessary to emphasize that the adiabatic temperature gradient is only obtained when parcels of air do not gain or lose heat to their environment. In practice, mixing of air parcels, energy interchange by radiation and the condensation of water vapour produce a change in total energy of the air parcel. The latter two effects are known as non-adiabatic processes or diabatic processes. Because of the diabatic processes, environmental temperature profiles are rarely adiabatic, though we shall see that the concept is still a useful one in understanding the vertical stability of the atmosphere. The average observed temperature gradient in the lowest 10 km of the atmosphere is $\sim 6 \cdot 5$ K km^{-1}, compared with an adiabatic temperature gradient of $9 \cdot 8$ K km^{-1}.

The adiabatic temperature gradient for the ocean can also be derived from equation 2.5, but the final expression is not as simple, because of the more complicated form of the equation of state. In this case equation 2.5 becomes

$$dQ = C_p\, dT - T\left(\frac{\partial \alpha}{\partial T}\right)_p dp$$

Assuming $dQ = 0$ and substituting for the hydrostatic equation it can be

Table 2.5. Adiabatic temperature gradients for planetary atmospheres.

	Earth	Mars	Venus	Jupiter
Adiabatic temperature gradient (K km^{-1})	$9 \cdot 8$	$4 \cdot 5$	$10 \cdot 7$	$20 \cdot 2$

shown that $dT/dz = -T\beta g/C_p$, where $\beta = (1/\alpha)(\partial\alpha/\partial T)_p$. β is the coefficient of heat expansion of sea water at constant pressure.

Unfortunately both β and C_p for sea water are dependent on pressure, temperature and salinity. At atmospheric pressure, standard ocean salinity of 35‰ and a temperature of 0°C, the adiabatic temperature gradient is $-0\cdot035$ K km^{-1}, whilst at a pressure of 500 atm (equivalent to 5000 m) the adiabatic temperature gradient is $-0\cdot1$ K km^{-1}. Thus, the adiabatic temperature gradient in the ocean is $\leqslant 1\%$ that of the atmospheric adiabatic temperature gradient. However, because of the small variation of temperature in the deep-water masses, the adiabatic temperature gradient has to be taken into account when considering water masses at different depths.

We have seen that the adiabatic effect of pressure on temperature is significant in both atmosphere and ocean, and therefore any vertical motions will produce an adiabatic temperature change in a parcel of fluid. Also we have indicated that radiation, mixing and water vapour condensation (in the atmosphere) will also produce significant changes in temperature of a parcel of air or sea water. It is useful, therefore, to

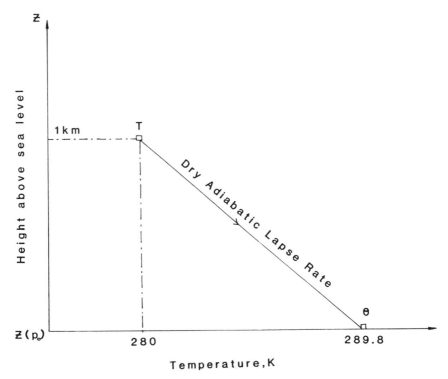

Figure 2.5. Relationship between temperature and potential temperature in the atmosphere as a function of height above mean sea level.

separate the adiabatic process from diabatic processes in order to distinguish the two effects. For instance, a meteorologist may wish to decide if the air mass at the surface is similar to the air mass observed in the upper atmosphere. Similarly, an oceanographer may wish to determine the similarity of water masses at two different depths in the ocean. For these processes it is necessary to remove the adiabatic effect, by bringing the air mass or water mass (adiabatically) to a reference pressure. This reference pressure is 10^5 Pa for the atmosphere, and surface atmospheric pressure for the ocean. Because of the small adiabatic effect in the ocean, the effect of variations in atmospheric pressure are

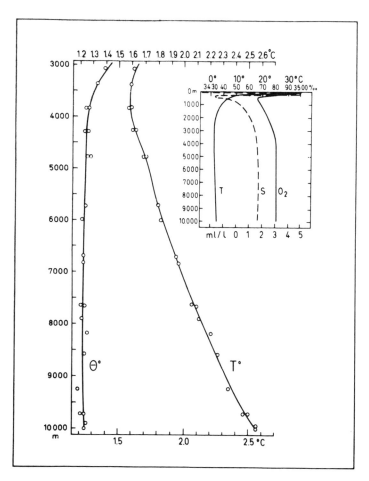

Figure 2.6. Observed temperature (T), *potential temperature* (θ), *salinity* (S), *and amount of oxygen* (O₂) *in the Philippines Deep (Mindanao Trench). Observations from the Danish vessel* Galathea *in July–August 1951. (Reproduced from* Descriptive Regional Oceanography, *by P. Tchernia, Pergamon Marine Series Vol. 3, p. 30, figure 3.2, with permission of Pergamon Press.)*

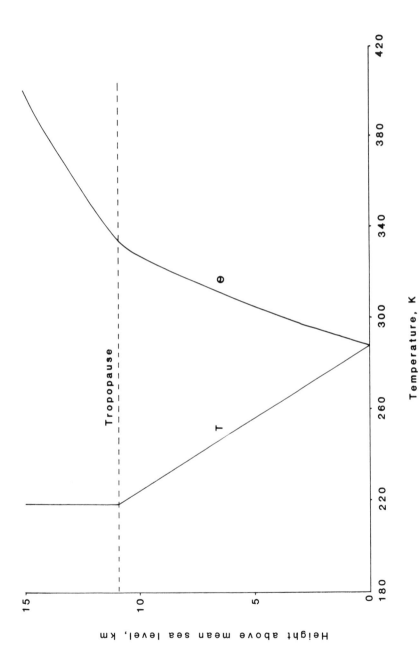

Figure 2.7. Mean vertical profile temperature (T) *and potential temperature* (θ) *in the lowest 15 km of the atmosphere.*

negligible. The temperature of the fluid parcel at the reference pressure is known as the potential temperature, θ. Figure 2.5 shows the temperature change of an air parcel moved adiabatically a vertical distance of 1 km from the level of the reference pressure. The temperature at the initial position is 280 K, but the potential temperature is 289·8 K. Generally, since the reference pressure is close to sea-level pressure, the atmosphere potential temperatures are usually larger than the observed temperatures. In the ocean, the reference pressure is the minimum pressure of the ocean (i.e. surface) and therefore ocean potential temperatures are always lower than *in situ* temperatures. Figure 2.6 shows the importance of reducing *in situ* temperatures to potential temperatures in the deep ocean. The measurements show temperature increasing steadily towards the bottom in the Mindanao Trench, which suggests a strong vertical instability of the lower layer (assuming salinity is not stabilizing the temperature gradient). However, the potential temperature is in fact nearly constant with depth, rather than increasing with depth. The apparent warm bottom layer is therefore solely an artefact of the adiabatic effect.

An example of the reduction of an observed atmospheric temperature profile to a profile of potential temperature is given in figure 2.7. It is again noted that the removal of the adiabatic influence substantially changes the profile, from one which decreases with height to a profile where the potential temperature increases with height. Usually, it is only close to the surface under strong heating conditions that a negative gradient of potential temperature exists. In the free atmosphere, generally potential temperature increases with height. The reason for this will be discussed in the following section.

2.6. *Vertical stability of the ocean and atmosphere*

We will first consider the stability of an incompressible fluid in the two situations shown in figure 2.8(*b*), (*d*). Initially the parcel of fluid at point 0 is in hydrostatic equilibrium with its environment (i.e. the upward pressure gradient on the parcel is balanced by the downward force due to gravity). If the parcel of fluid in case (*b*) is moved *upwards* by an external force to a position *A*, and provided there is no mixing of the parcel with its environment, then the parcel will have a density greater than its environment density at that level. The parcel is now heavier than its environment, and will therefore accelerate back towards its original level. In this case we say that the fluid system is *stable*. However, in case (*d*), the parcel at *A* will have a density less than the environment density at that level. In this case the fluid is lighter than its surroundings and the parcel will be subjected to an upward buoyancy force, which will accelerate the

STABLE

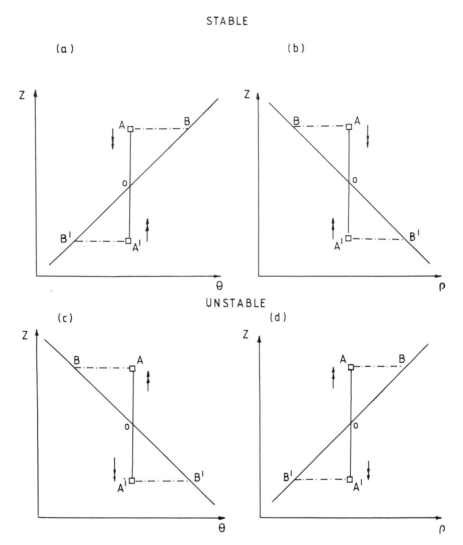

Figure 2.8. Stability of a parcel perturbed from equilibrium for atmosphere (a, c) *and ocean* (b, d). *For the purposes of illustration it is assumed that the ocean is incompressible.*

parcel away from its original position. Furthermore, as the parcel moves further from its initial position it will be subjected to a larger buoyancy force and therefore will be subjected to larger accelerations. In this case the fluid system is *unstable*. It should be appreciated that downward displacements of the parcels from the initial position will behave in a similar manner. In the stable case the fluid parcel will move back to its original position, whilst in the unstable case it will accelerate downwards

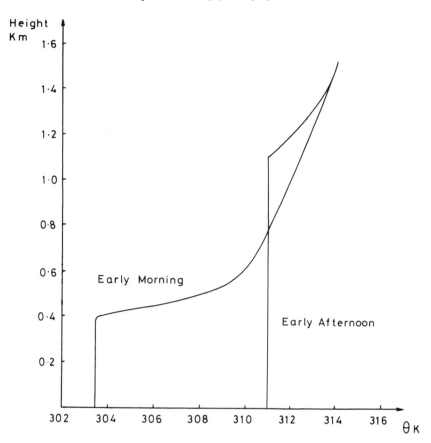

Figure 2.9. Diurnal variation in potential temperature in the atmospheric boundary layer over land in summer. (Reproduced from R. B. Stull (1973), J.A.S., 30, 1097, figure 5, with permission of the American Meteorological Society.)

and away from its original position. In the situation where the density is uniform throughout the fluid, there will be no buoyancy forces and the system is said to be neutral. In this case, parcels will continue to move in their initial direction until frictional forces slow them down.

The above criterion for stability can be applied to both ocean and atmosphere provided the adiabatic effects are taken into account. We have shown that a conservative property of an adiabatic fluid is the potential temperature and in the atmosphere the criterion for stability is simply based on the vertical gradient of potential temperature. If the potential temperature *increases* with height, the atmosphere is *stable* (figure 2.8(*a*)), whilst if it *decreases* with height it is *unstable* (figure 2.8(*c*)). It is left to the reader to consider the buoyancy forces on parcels displaced from equilibrium for these three cases, i.e.

$$\text{stable} \qquad \frac{\partial \theta}{\partial z} > 0$$

$$\text{neutral} \qquad \frac{\partial \theta}{\partial z} = 0 \qquad\qquad (2.7)$$

$$\text{unstable} \qquad \frac{\partial \theta}{\partial z} < 0$$

Figure 2.9 shows the diurnal variation in potential temperature in the lowest 1 km of the atmosphere over land. During the day-time, the potential temperature is uniform up to the height of the convection and the potential temperature gradient is stable above this level. Only very close to the surface can a negative gradient of potential temperature exist. The reason for this is that the radiational heating of the surface can cause a higher surface temperature than air temperature and sustain a negative gradient close to the surface. However, away from the surface large convection currents driven by the unstable temperature gradient can mix the boundary layer very effectively and thus produce a uniform potential temperature (i.e. $\partial \theta / \partial z = 0$) is produced. As midday approaches the surface heating reaches a maximum, and the convection currents become deeper and mix stable air into this boundary layer. Therefore the region of uniform potential temperature (or the mixed layer) increases in depth. During the night, the ground cools by loss of planetary radiation more effectively than the air above it and therefore the potential temperature gradients becomes positive, and the air becomes stable. In those circumstances, the air is quiescent and smoke and other pollutants will tend to be trapped close to the ground. Only by artifically adding buoyancy to the pollutants can they be dispersed effectively.

In the ocean, the situation is complicated by the effect of salinity, hence the vertical gradient of potential temperature does not always indicate the stability of the ocean. Generally, we have to use a quantity called the potential density (σ_θ), which is the density a parcel of sea water would have if it was moved adiabatically to the sea surface.†

The stability of the ocean is then determined by the vertical gradient of potential density, i.e.

$$\text{stable} \qquad \frac{\partial \sigma_\theta}{\partial z} < 0$$

$$\text{neutral} \qquad \frac{\partial \sigma_\theta}{\partial z} = 0 \qquad\qquad (2.8)$$

$$\text{unstable} \qquad \frac{\partial \sigma_\theta}{\partial z} > 0$$

---

†For deep water masses a reference level of 4 km is used.

Unstable density gradients are very rarely observed in the ocean. The reason for this is that the convective instability process is very efficient in mixing the unstable density gradient to produce a neutral profile. The presence therefore of a well-mixed ocean layer is a good indication of a convective instability process. The principal cause of the convective instability in the ocean is strong cooling of the surface during winter, which causes the formation of a relatively dense surface layer which will be unstable with respect to the deeper water and cause overturning. The surface cooling has a dual effect in destabilizing the water. It will not only remove heat from the water column but increased evaporation will tend to increase the surface salinity and thus increase the density. The depth to which convection occurs will depend on the vertical density structure and the intensity of the surface cooling. In the North Pacific, the depth of winter convection is limited to the upper 150 m, because of the presence of a very stable halocline below this depth. However, in the Labrador Sea deep convection has been observed to occur to a depth of 1500 m, and in the western Mediterranean in the Gulf of Lyons deep convection occurs to the bottom.

Buoyancy frequency

Let us reconsider the way in which a parcel of fluid returns to its equilibrium position (figure 2.8) in a stable system. If there is no mixing of the parcel with its environment and no frictional drag on the parcel, then the parcel will accelerate towards its equilibrium position and overshoot. It will then find itself in an environment of higher density, and it will accelerate in the opposite direction back towards its equilibrium position. The parcel will therefore oscillate about its equilibrium position with a frequency related to the magnitude of the density gradient. If the density gradient is large, then the buoyancy accelerations will be correspondingly increased, and the frequency will also increase. This frequency is called the Brunt–Väisälä frequency or the buoyancy frequency (N).

Consider the buoyancy force on a parcel displaced a distance δz from its equilibrium density ϱ_e. The buoyancy force is $-(\varrho_e - \varrho)g$. If $\varrho = \varrho_e + (\partial \varrho/\partial z)\,\delta z$ then $(\varrho_e - \varrho)g = -g(\partial \varrho/\partial z)\,\delta z$. The acceleration as a result of the buoyancy force is $dw/dt = g/\varrho\,(\partial \varrho/\partial z)\,\delta z$. If $w = d\,\delta z/dt$ then

$$\frac{d^2}{dt^2}\delta z = -\left(-g\frac{1}{\varrho}\frac{\partial \varrho}{\partial z}\right)\delta z$$

This equation is analogous to the simple harmonic equation $d^2x/dt^2 = -N^2 x$, where N is the frequency. Hence $N = \sqrt{[-(g/\varrho)(\partial \varrho/\partial z)]}$. Applying this expression to the atmosphere, the density is replaced by potential temperature, i.e.

$$N = \sqrt{\left(\frac{g}{\theta}\frac{\partial\theta}{\partial z}\right)} \qquad \text{where} \quad \frac{\partial\theta}{\partial z} \geqslant 0$$

for the ocean   (2.9)

$$N = \sqrt{\left(\frac{-g}{\sigma_\theta}\frac{\partial\sigma_\theta}{\partial z}\right)} \qquad \text{where} \quad \frac{\partial\sigma_\theta}{\partial z} \leqslant 0$$

where σ_θ is the potential density.

The buoyancy frequency N provides a useful way of describing the stability of both the atmosphere and the ocean, and it is a very important parameter in the advanced study of dynamical oceanography and meteorology. One of its most important uses is the determination of the frequency of internal waves, which are continually produced by turbulent motions in both systems. In the ocean thermocline N has a period of 10 min whilst in the deep ocean, where the stability is weaker, it has a typical period of 50 hours. Internal waves in the thermocline, produced by surface storms, can have amplitudes of 10–50 m. In the atmosphere, the period of the waves is shorter than that of waves in the ocean, ranging from 30 s to 10 min. The more spectacular internal waves are produced by forced ascent over mountains, which in certain circumstances can produce wave trains (lee waves) extending for up to 100 km downstream of the mountain. The waves are most apparent when the wave crests are above the condensation level.

3

Radiation, temperature and stability

3.1. Vertical variation of atmospheric constituents

The vertical variations of constituents in the atmosphere are important in understanding the way in which solar and planetary radiations are absorbed in the atmosphere, and the resulting effect on the temperature profile of the atmosphere. Here the physical processes which control the distribution of atoms and molecules in the atmosphere will be considered. First, consider the major constituents of the atmosphere: oxygen, nitrogen, argon, water vapour and carbon dioxide. If all these constituents were initially mixed together and left for a long time in an undisturbed environment, the distribution of each constituent would vary with height according to its molecular or atomic weight. The lighter molecules would have a smaller variation with height than the heavier molecules and thus the lighter molecules would be more abundant in the upper atmosphere than in the lower atmosphere compared with the heavier molecules. The equilbrium distribution is given by the hydrostatic equation for each constituent gas rather than for the mixture of gases. Therefore it can be seen that the rate of change of partial pressure (P_i) of each gas is directly proportional to the molecular weight (m_i) of the gas, i.e.

$$\frac{\partial P_i}{\partial z} = - \frac{m_i P_i g}{R^* T}$$

However, the observed distribution of the gases in the atmosphere does not show this relationship. All of the gases, even carbon dioxide which is the heaviest, are well mixed to a height of 100 km and it is only above this level that evidence of gravitational settling is apparent. To understand this distribution it is necessary to consider the rate at which diffusion occurs. The rate of gravitational settling of gases is determined by the molecular diffusion rate for each gas. The diffusion rate is defined by the product of the velocity of the molecules and their mean free path

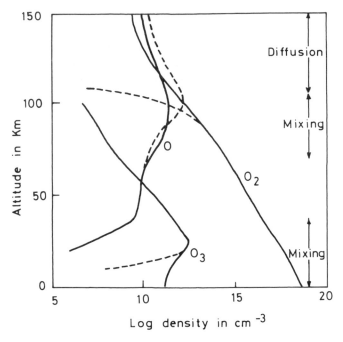

Figure 3.1. Altitude profiles of the densities of atomic oxygen (O), molecular oxygen (O₂) and ozone (O₃) in the Earth's atmosphere. Broken lines indicate theoretical profile crossing photochemical equilibrium. (Reproduced from Atmospheres, *by R. M. Goody and J. C. G. Walker, p. 25, figure 2.2, with permission of Prentice-Hall.)*

between collisions. The velocity of the atom or molecule increases with the temperature of the gas and decreases with the molecular weight. Hence, the diffusion rate is higher for light gases at low pressures and high temperatures than for heavier gases at low temperatures and high pressures. There is a rapid decrease of pressure with height in the atmosphere and because of this the most significant factor in the determination of diffusion rates is the mean free path. The diffusion rate is many orders of magnitude higher in the upper atmosphere, where the mean free path is very large, than in the lower atmosphere. Typically at a height of 120 km it would take approximately 1 day for gravitational separation to occur, but for the atmosphere taken as a whole the typical time scale is approximately 30,000 years. The atmosphere is always in a state of turbulent motion and therefore the molecules are being continually mixed by a variety of atmospheric motions. The rates of mixing are much larger than the corresponding rates of gravitational settling in the lower atmosphere, and therefore the profiles of these constituents are well mixed. Only when the gravitational settling time becomes of the same order as the mixing time scale is diffusive separation observed. For molecular oxygen and nitrogen this occurs typically at a height of 100 km. Above these levels the gases are distributed according to their molecular weights, and therefore hydrogen and helium occur in increasing abundancies at high levels, whilst the abundancies of the heavier gases, such as oxygen and nitrogen, decrease with height.

It is only for the noble gases that the story finishes here. For all the other constituents, their interaction with radiation has to be considered when discussing their vertical distribution in the atmosphere. These interactive processes include:

(i) The dissociation of a molecule by radiation.
(ii) The combination of atoms by collisions.

The dissociation energy of a molecule is the energy required to break the molecular bond. The energy of one photon of radiation is given by $h2\pi/\lambda$, where h is Planck's constant and λ is the wavelength of the photon, and it is therefore accepted practice to define the dissociation energy in terms of the wavelength of the radiation. For an oxygen molecule the dissociation energy corresponds to a wavelength of $0 \cdot 24 \ \mu m$, and therefore all radiation having a smaller wavelength, and hence a higher energy, than this will dissociate the oxygen molecules into two atoms of oxygen, i.e.

$$O_2 + h\nu \rightarrow O + O, \qquad \lambda \leqslant 0 \cdot 24 \ \mu m$$

Nitrogen has a higher binding energy than oxygen and only photons having wavelengths less than $0 \cdot 13 \ \mu m$ are able to dissociate the nitrogen molecule.

Atomic oxygen becomes an increasingly abundant constituent above a height of 20 km, but it is only above 120 km that molecular oxygen is replaced by atomic oxygen as a dominant oxygen constituent. The reason for this distribution is that the rate of molecular dissociation is proportional to the product of the number of oxygen molecules and the number of photons having more than the required energy for dissociation. At very high elevations, there are a large number of photons with the energy necessary to dissociate oxygen molecules but, due to gravitational settling and dissociation, there are few oxygen molecules and therefore the rate of production of atomic oxygen is low. However, in the lower atmosphere, though there is an abundance of molecular oxygen, there are only a few photons possessing the required energy to dissociate the molecules because many of these photons have been absorbed in the upper levels of the atmosphere. Therefore, it is only in the middle levels, around 100 km, where there are sufficient molecules and photons to produce a large dissociation rate and hence to produce a high number of oxygen atoms. However, the predicted height of maximum atomic oxygen production does not agree with the observed distribution. This is due to two additional processes in the lower atmosphere which have so far been neglected and which act to reduce the number of oxygen atoms. The first process is the recombination of atomic oxygen into molecular oxygen by collision. This process is very effective at higher pressures and it is therefore a major sink of atomic oxygen at lower altitudes, between 70 and 100 km. The second process is the production of ozone, O_3, from a three-way collision between an atom of oxygen, a molecule of oxygen and another unspecified molecule, M:

$$O + O_2 + M \rightarrow O_3 + M$$

The unspecified molecule is necessary because it absorbs the energy released when a molecule of ozone is formed. The production of ozone depends on the relative numbers of oxygen molecules and atoms and the number of three-way collisions, which depends in turn, on the total pressure of the gas. For a photochemical equilibrium, the rate of production of atomic oxygen must be equal to its rate of destruction, and given the photoabsorption properties of the constituents, it is possible to obtain the vertical equilibrium distribution. This profile is known as the Chapman profile.

Figure 3.1 shows the relative distributions of atomic oxygen, molecular oxygen and ozone in the atmosphere. It is noted that because the ozone distribution is dependent on both the collision frequency and the amount of atomic oxygen, it tends to have its highest production rate between 30 and 100 km. Below 30 km the infrequency of occurrence of atomic oxygen reduces the rate of ozone formation to insignificant levels. Above 100 km it is the low collision frequency which brings about the reduction of the rate of ozone formation. However, there are three major processes which result in the destruction of ozone:

(i) The photodissociation of ozone which occurs for photon wavelengths less than 11 μm, which therefore operates throughout the solar and planetary spectra

$$O_3 + h\nu \rightarrow O_2 + O, \qquad \lambda \leqslant 11 \ \mu m$$

The strongest dissociation occurs for wavelengths between $0 \cdot 2$ and $0 \cdot 3$ μm, and therefore ozone is a very effective absorber of ultraviolet radiation.

(ii) The recombination of ozone with atomic oxygen

$$O_3 + O \rightarrow 2O_2$$

(iii) The removal of ozone at the Earth's surface by oxidation.

It can be seen that, at high levels of the stratosphere, ozone will be readily destroyed by processes (i) and (ii), whilst process (iii) will destroy ozone in the troposphere. In the troposphere mixing rates are larger than in the stratosphere, where most ozone is formed, and therefore ozone molecules entering the troposphere are brought to the surface in a matter of a few days and oxidation then takes place. Hence ozone tends to have a concentration maximum in the lower stratosphere, where destructive processes are at a minimum. Here, individual ozone molecules may have lifetimes of the order of a few months. The detailed photochemistry of ozone is further complicated by other trace gases, such as nitrous oxide (which is related to biological production processes at the surface) and chlorofluoromethanes (present as aerosol propellants in many sprays) and other man-made gases. All of these gases, in certain circumstances, can act to reduce the level of ozone in the atmosphere and hence reduce the ultraviolet shielding at present enjoyed by the Earth. However, quantitative evaluations are uncertain, because of the large number of possible photochemical reactions and many other factors, such as dynamical motions in the stratosphere.

The inter-relationship between atomic oxygen, molecular oxygen, ozone and radiation is an example of the complexity of photochemical processes which produce the observed distributions of the constituent gases of the atmosphere. For the other principal gases, water vapour and carbon dioxide, dissociation is only significant at altitudes greater than 100 km. Carbon dioxide is dissociated by ultraviolet radiation of wavelengths less than $0 \cdot 17 \ \mu m$ and water molecules commence dissociation at $0 \cdot 24 \ \mu m$. The vertical distribution of water vapour in the atmosphere below 100 km is determined principally by the condensation process. The rapid decrease of pressure and temperature with height in the lowest 20 km of the atmosphere results in a reduction of the saturated mixing ratio (defined as the ratio of the mass of water vapour in a given volume of air, saturated with respect to a plane water surface, to the mass of dry air in the same volume) from 10^{-2} at the surface to 10^{-5} at 20 km. Therefore for every tonne of dry air in the stratosphere there is only 10 g of water vapour. The temperature minimum at the tropopause, 10 km above the surface, acts as a cold trap for the movement of water vapour between the troposphere and the stratosphere. However, the water vapour in the upper atmosphere is still an important constituent when considering the absorption of solar and planetary radiation. At 80 km, where the temperature reaches a second minimum at the mesopause, occasionally there is sufficient water vapour to allow the formation of thin, noctilucent ice clouds. These clouds cannot be observed during the day but are often seen just after sunset in high latitudes, when the reflected sunlight from the clouds may be observed at the surface.

The reasons for the temperature minima at the tropopause and mesopause will be discussed in Section 3.4.

3.2. *The attenuation of solar radiation*

The observation that the maximum in the solar-radiation spectrum occurs at $0 \cdot 5$ μm, whilst planetary radiation peaks between 10 and 15 μm (in the middle infra-red), allows separation of the radiation processes due to solar-radiation absorption from those involved in planetary-radiation absorption. Only $0 \cdot 4\%$ of the total solar radiation has a wavelength longer than 5 μm, whilst a similar small percentage of planetary radiation from a surface at a temperature of 288 K has a wavelength less than 5 μm. The propagation of solar radiation through the atmosphere is simpler to understand than that of planetary radiation, because no re-emission of radiation occurs as the solar radiation travels through the atmosphere and therefore it is only necessary to consider the attenuation factors. The situation with respect to the planetary-radiation spectrum is complicated by the re-emission of radiation at similar wavelengths to the absorbed radiation. This will be considered in Section 3.3.

Figure 3.2 shows the various factors leading to the attenuation of solar radiation in a typical atmosphere. In a dust-free and pollution-free atmosphere with no clouds, the attenuation would be the result of two processes:

(i) Gaseous absorption.
(ii) Molecular scattering.

The primary gaseous absorption occurs in the ultraviolet region of the spectrum, at wavelengths less than $0 \cdot 3$ μm. As previously discussed, this absorption mainly results in the photodissociation of molecular oxygen and nitrogen above 100 km and the photochemical reactions involving ozone in the stratosphere. This primary gaseous absorption results in a nearly complete removal of lethal ultraviolet radiation by the upper atmosphere. However, a small fraction of the incident ultraviolet radiation does manage to reach the surface, particularly when the sky is clear, and this can be damaging to human skin and, with excessive exposure, can cause skin cancers. In quantitative terms, the amount of ultraviolet radiation absorbed is less than 2% of the total solar flux, for wavelengths less than $0 \cdot 32$ μm. Ozone has absorption bands in the visible region of the spectrum but these bands are generally weaker than the ultraviolet absorption bands. The only other significant absorber of solar radiation is water vapour, which has a number of absorption bands in the near infra-red part of the solar-radiation spectrum. The absorption bands of water vapour have a very significant quantitative effect on the amount of solar radiation actually reaching the surface, because of the large amount of solar radiation in this region of the spectrum, and this effect is illustrated in figure 3.2. The variability of the amount of water vapour present in the atmosphere means that the effect of its infra-red absorption will vary over the globe.

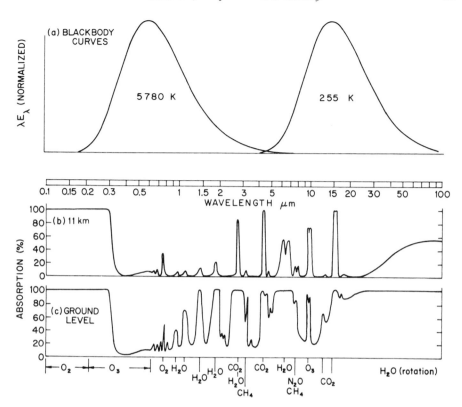

Figure 3.2. Attenuation of solar radiation. (a) Normalized blackbody curves for 5780 and 255 K, plotted so that irradiance is proportional to the areas under the curves. (c) Atmospheric absorption in clear air for solar radiation with zenith angle of 50° and for diffuse terrestrial radiation. (b) Same as for (c) but for the portion of the atmosphere lying above the 11 km level, near the middle-latitude tropopause. (Adapted from Atmospheric Radiation *by R. M. Goody, p. 4, figure 1.1, with permission of Oxford University Press.)*

Molecular scattering is proportional to λ^{-4}, where λ is the wavelength of the incident photon. Thus the shorter-wavelength light is scattered rather more than the longer-wavelength light. For example, blue-light scattering is four times greater than red-light scattering and hence the direct solar beam loses a significant fraction of its blue light. On a clear day, with a dust-free atmosphere, the shortest wavelength radiation is seen at the antisolar point, i.e. the point at 180° to the direction of the Sun and at the same elevation as the Sun. This part of the sky often has a purple hue and ultraviolet radiation is a maximum from this direction.

The presence of small suspended particles, of diameter generally less than a few microns, in the atmosphere has a significant influence on the solar radiation received at the surface. Such particles, called the aerosol, can both scatter and absorb radiation. The scattering of solar radiation tends to increase the path length of a photon through the atmosphere and therefore it leads, indirectly, to an increase in the probability of absorption. The scattering by the aerosol varies as λ^{-1}, i.e. it is not so strongly wavelength dependent as molecular scattering, and therefore it is significant throughout the solar spectrum. From figure 3.3 it may be seen that the attenuation by the aerosol increases as the elevation of the Sun decreases. This is the result of the increased path length through the atmosphere and, therefore, the increased chance that a photon will be absorbed. Hence, in polar latitudes in summer, an aerosol layer will tend to produce enhanced absorption.

There are many types of aerosol in the atmosphere, but a broad classification into major types is possible. Considering the Earth as a whole, the natural aerosol dominates over the man-made aerosol, although the latter is dominant in highly industrialized regions such as Europe and the eastern U.S.A. The natural aerosol is produced by wind turbulence at the Earth's surface. Over land the particles are mainly silica

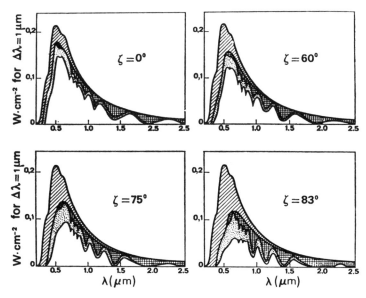

Figure 3.3. Molecular scattering, absorption and scattering by aerosols. In each of the four diagrams (corresponding to four values of the zenith angle, ζ, of the Sun) the upper curve shows the spectral distribution of the solar constant, while the lower curve represents the spectral distribution of the direct solar beam at the sea surface, for clear dry weather (equivalent to water thickness equal to 1 cm). The areas of oblique lines, cross hatching and dots represent, respectively, the effects of molecular scattering, absorption and scattering by aerosols. (Reproduced from Ch. de Brichambault and G. Lamboley (1968), Cahiers de L'A.F.E.D.E.S., No. 1, pp. 11–109.)

and clay and over the ocean they are mainly salt. In addition, volcanic activity can inject large quantities of sulphur particles into the troposphere and the lower stratosphere. In the El Chichon eruption in 1982 it is estimated that 10^7 tonnes of material was injected into the stratosphere and, in Hawaii, the direct solar radiation was reduced by approximately 7% following the explosion. The aerosol produced by El Chichon was traced across the Pacific Ocean for many months. Other sources of the natural aerosol include forest fires, which can inject large quantities of carbon particles into the troposphere. These carbon particles are particularly effective in absorbing solar radiation. An important type of aerosol is formed from hygroscopic substances, such as sodium chloride. These aerosols act as very effective condensation nuclei and hence maritime salt haze becomes particularly effective in reducing solar radiation when the humidity reaches approximately 95%.

Human activities increase the input of sulphur into the troposphere, in the form of sulphur dioxide and ammonium sulphates, and the effect of this input is important in the formation of acidic rain, although its influence on radiation is probably small. However, the carbon input, from coal- or oil-based industries, can produce very significant reductions in the solar radiation. Again, as with the salt haze, as the humidity increases condensation occurs on hygroscopic pollutants, such as ammonium sulphate, and this increases their influence on the solar radiation.

Figure 3.4 shows how the aerosol may affect the path of a photon through the atmosphere. The back-scattering of incident photons from the aerosol will tend to enhance the albedo of the atmosphere. The photons which are not back-scattered into space, in the absence of gas absorption or clouds, will be either absorbed by the aerosol or reach the Earth's surface. Some of the photons will be reflected back into the atmosphere by the surface. A fraction of the reflected photons will be back-scattered to the surface by the aerosol and the remaining photons will be lost to space. The relative magnitudes of each fraction will depend on the absorption to back-scattering ratio of the aerosol and the surface albedo. Although these properties depend on the type of aerosol and the type of surface, some broad trends have been identified. First, if the surface albedo is less than 30%, then the aerosol will tend to increase the planetary albedo. Surface albedos of such magnitudes are typical of all ocean surfaces and most land surfaces, except for sand deserts. If the surface has a higher albedo, light trapping between the surface and the aerosol will tend to decrease the planetary albedo. Secondly, the absorption to back-scattering ratio varies between 1% and 20% for maritime 'salt particles' and rises to 500% for carbonaceous particles. Thus maritime aerosols attenuate radiation primarily by scattering whilst carbonaceous particles attenuate radiation by absorption. Figure 3.5 shows the effect on solar radiation of a dust storm in the southern Sahara. It is

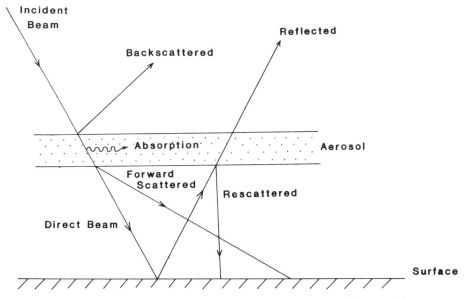

Figure 3.4. Schematic diagram of scattering and absorption by an atmospheric aerosol layer.

noticed that the direct solar radiation is dramatically reduced by the dust, although the global solar radiation, i.e. the radiation from all directions, which includes the scattered radiation, is only reduced by approximately 15%. Hence, the major influence of the dust is the scattering of the light field rather than the absorption of radiation.

The last and most significant influence on solar radiation is that of clouds. Cloud droplets are typically larger than aerosol particles, usually varying between 1 and 20 μm in diameter. In this size range, the scattering is dominated by diffraction which tends to produce scattering at small angles from the angle of incidence. This diffraction-dominated scattering does not tend to be dependent on wavelength. The absorption to backscattering ratio for clouds is very small and therefore the predominant influence on solar radiation is the back reflection. Particularly spectacular is the reflection from cumulus clouds on an otherwise clear day. Of course, ice is an important scatterer of radiation as well, but it tends to be more effective at scattering radiation at larger angles from the angle of incidence than water droplets and therefore, close to the Sun, ice clouds appear less bright than water clouds. However, the opposite is the case for clouds in the direction away from the Sun. In addition, because the ice crystals are hexagonal, plate and columnar prism optical processes are important and these produce a fascinating series of optical phenomena.

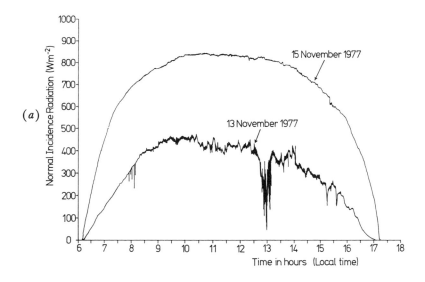

(a)

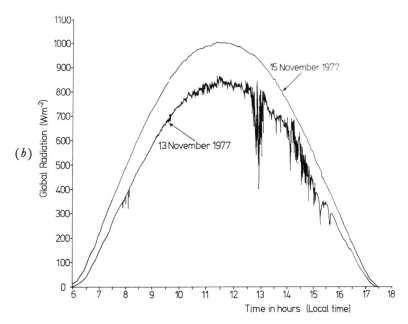

(b)

*Figure 3.5.   Solar radiation attenuation during a dust storm. (a) Normal incidence of radiation during Harmattan (13 November 1977) and clear sky (15 November 1977) as a function of local time. (b) Same as (a) except for global radiation (direct + diffuse). Note the large depletion of direct solar radiation but the smaller depletion of global radiation. (Reproduced from J. Adetunji et al. (1979), Weather, **30**, 434 and 435, figures 4 and 5, with permission of the Royal Meteorological Society.)*

The more common phenomena include the appearance of a 22° halo around the Sun and the production of parhelia, or mock Suns.

The quantitative influence of clouds on solar radiation is dependent on:

(i) The solar elevation.
(ii) The thickness of the cloud.
(iii) The size, distribution and amount of liquid water in the cloud.

Though absorption is small, typically less than 10% and probably often as low as 5%, it is enhanced by scattering, which increases the path length within the cloud and hence the probability of water-vapour absorption. The transmission of the direct beam is also dependent on the type of cloud. For example, the direct beam is significantly attenuated by low-level clouds and therefore a stratocumulus cloud of 400 m thickness will be sufficient to obliterate the Sun's disc. On the other hand, cirrus clouds allow the Sun's disc to be seen. As the thickness of the cloud increases, the transmission decreases and the albedo increases and approaches a value between 80 and 90%. A decrease in the elevation of the Sun will also result in a decrease in the transmission of the direct beam and an increase in back-scattering by clouds. A generalized quantitative analysis is not possible, however, because of the differences between clouds of the same general type.

Figure 3.6 shows a detailed breakdown of the downward and upward components of solar radiation measured through low-level stratocumulus cloud, which is particularly common over the ocean. The cloud is 200−400 m thick, on average, and, in common with other low-level clouds, it has a high liquid water content of between $0 \cdot 1$ and $0 \cdot 5$ g m^{-3}. The albedo of the cloud is 68% and, of the solar radiation not reflected, 25% is transmitted through to the surface and 7% is absorbed by the cloud. Though the absorption is a relatively small contribution to the attenuation of the incident solar radiation when compared with the amount reflected, the absorption is equivalent to 50 W m^{-2}, which is sufficient to warm the upper part of the cloud by 20−40 K day^{-1}, and offset the long-wave cooling from the cloud top. It is noted that only a small fraction of the downward solar radiation is reflected from the surface and this is consistent with the general observation of low albedo over ocean surfaces.

For a non-uniform broken cloud, e.g. cumulus, it is more difficult to obtain quantitative measurements. Figure 3.7(a) shows the variation of downward solar radiation in a broken stratocumulus layer over a distance of 85 km. The large variability in solar radiation between clear patches and cloud is clearly evident, with changes of 50% of the average value. However, in addition to this, the downward solar radiation in clear patches is higher than would be observed under clear skies. This is caused

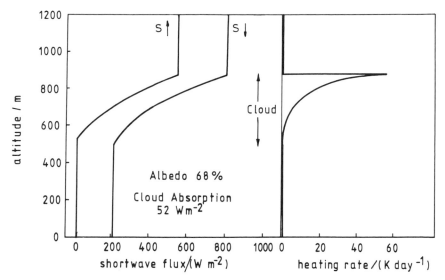

Figure 3.6. The observed short-wave (S) fluxes measured from three aircraft during the JASIN experiment (August 1978), corrected to a solar zenith angle of 43·7°, for a stratocumulus cloud. The calculated heating rate profile is shown in the right-hand diagram. (Reproduced from Schmetz et al. (1983) Phil. Trans. Roy. Soc., **A308**, *380, figure 2, with permission of the Royal Society.)*

by the side reflection from the surrounding clouds acting to enhance the downward radiation. For broken cloud it is necessary, therefore, to obtain statistical relationships between the fraction of cloud cover and the transmission. An example of a relationship is shown in figure 3.7(*b*) for stratocumulus cloud. In this respect, satellite techniques can be particularly powerful in estimating albedos for ensembles of clouds.

3.3. *Absorption of planetary radiation*

The predominant absorption of planetary radiation is by water vapour, carbon dioxide and ozone. All of these gases consist of triatomic molecules which have vibration–rotation absorption bands in the infrared region of the spectrum, unlike diatomic molecules, such as oxygen and nitrogen, and the noble gases which exist as single atoms.

The detailed properties of the absorption curves are complicated, particularly for water vapour as illustrated by figure 3.2, but some bands are much more intense than others. These strong bands occur in the near infra-red at 1·4, 1·9, 2·7 and 6·3 μm for water vapour and the continuum absorption band stretches from 13 to 1000 μm. The importance of this latter absorption band cannot be overemphasized, since more than 50% of the planetary radiation occurs at wavelengths greater

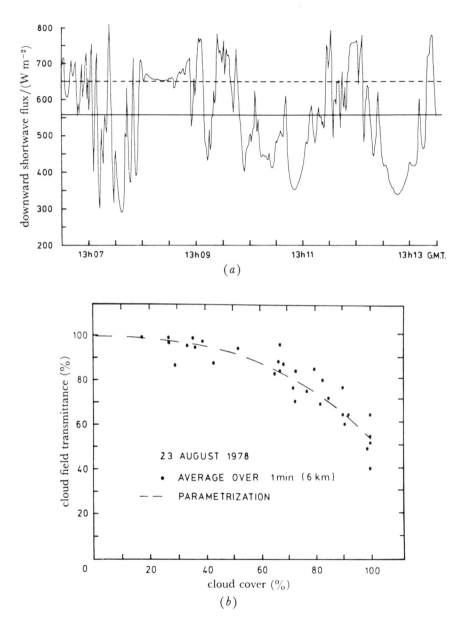

Figure 3.7. (a) Downward short-wave flux as measured under a broken stratocumulus layer on 23 August 1978. The dashed line is the calculated clear sky flux and the solid line the average flux. (b) Transmission of a broken stratocumulus layer as a function of cloud amount. (Reproduced from Schmetz et al. (1983), Phil. Trans. Roy. Soc., **A308**, 382 and 383, figures 5 and 6, with permission of the Royal Society.)

than 13 μm. Carbon dioxide has its principal absorption bands at $2 \cdot 7$ and $4 \cdot 3$ μm with a further strong band at $14 \cdot 7$ μm. Ozone has bands at $4 \cdot 7$, $9 \cdot 6$ and $14 \cdot 1$ μm. The $9 \cdot 6$ μm absorption band coincides with the maximum of the planetary radiation emitted from the surface. One of the broad features of the absorption curve typical of the atmosphere as a whole is the absence of absorption between 8 and 13 μm, except for the $9 \cdot 6$ μm absorption of ozone. It is noted that this region between 8 and 13 μm is within the peak emission region of the planetary radiation spectrum and some 32% of the total planetary emisison resides within it. The region is known as the 'atmospheric window' because, in the absence of clouds, planetary radiation should pass through the atmosphere almost unimpeded. However, recent work has shown that, with high water-vapour concentrations typical of tropical latitudes, additional absorption peaks appear in the window region as the result of the formation of water-vapour dimer, $(H_2O)_2$. The water dimer is formed by the joining of two water molecules by hydrogen bonding. Although it exists only in low concentrations, when compared with water vapour, it has significant rotation–vibration absorption bands in the atmospheric window. With very high water-vapour concentrations, typically greater than 20 mb in tropical environments, the dimer has an important influence in reducing the long-wave radiation lost to space. The presence of this dimer may lead to doubling of the absorption at 10 μm (under conditions favourable to its formation). Other minor gases, such as nitrous oxide, methane and carbon monoxide, also have absorption peaks in the infra-red spectrum and therefore contribute to the absorption of planetary radiation.

In addition to gaseous absorption, the continuous absorption throughout the infra-red spectrum of liquid water droplets is also of prime importance in the absorption of planetary radiation. At a wavelength of 10 μm a liquid water film only 60 μm thick is sufficient to absorb 95% of the incident radiation, although care has to be exerted to distinguish between the absorption by plane water surfaces and by spherical water droplets. In the latter case, scattering becomes an important process. As an example, a cloud 200 m thick with a liquid water concentration of $0 \cdot 1$ $g\,m^{-3}$ would absorb about 90% of the incident planetary radiation. Therefore a reasonable thickness of unbroken cloud would be sufficient to absorb all the incident planetary radiation. Ice has a similar infra-red absorption to liquid water, but the total water content in ice clouds is generally much less than that found in mixed and water droplet clouds lower in the troposphere because of the low temperatures which prevail. Hence, ice clouds will not be as effective in absorbing infra-red radiation. Typically, cirrus cloud has an infra-red absorptivity of 35%, compared with an absorptivity of 100% for stratocumulus.

Additional absorption by aerosols occurs, but this effect is not as marked in the infra-red as it is at visible wavelengths, because of the λ^{-1}

dependence. However, the importance of hygroscopic nuclei when considering infra-red absorption cannot be overstressed, since the particles will grow by condensation and absorption will increase because of the water which becomes attached to the nuclei. It should therefore be expected that a significant increase in absorption of planetary radiation will occur as the humidity rises and such an increase could well be anticipated in the marine atmosphere and over industrial areas where hygroscopic nuclei may be produced in large concentrations.

3.4. Vertical temperature profile and its relation to radiation

The mean vertical temperature profile for the atmosphere is shown in figure 3.8. The most important features are the three turning points, known as the tropopause, the stratopause and the mesopause, which allow a quantitative division of the atmosphere into four distinct layers; the troposphere, the stratosphere, the mesosphere and the thermosphere. The troposphere accounts for 80% of the mass of the atmosphere and it

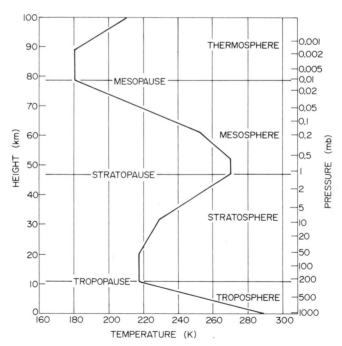

Figure 3.8. Vertical temperature profile for US Standard Atmosphere. (Reproduced from Atmospheric Science, *by J. M. Wallace and P. V. Hobbs, p. 23, figure 1.8, with permission of Academic Press.)*

is a region of strong vertical mixing. It is associated with our experience of 'weather' and is most important in the condensation of water vapour, the average decrease of temperature with height being approximately $6 \cdot 5 \, \mathrm{K \, km^{-1}}$. The stratosphere contains about $19 \cdot 9\%$ of the mass of the atmosphere and it is a region of very pronounced stability. The temperature increases with height at a rate of approximately $2 \, \mathrm{K \, km^{-1}}$, compared with a *decrease* of $9 \cdot 8 \, \mathrm{K \, km^{-1}}$ expected for a dry neutrally buoyant atmosphere. The stability of the stratosphere is most marked between 30 and 50 km, where the absorption of solar radiation by ozone reaches a maximum. It is noted that the maximum ozone concentration occurs in the lower stratosphere. Although the tropopause forms a pronounced stability barrier between the stratosphere and the troposphere, it is highly variable in position. Occasionally, the tropopause may extend down to 6 km and, in these conditions, stratospheric air may be drawn into the troposphere, most commonly along frontal boundaries. There is a high ozone concentration in this stratospheric air and it is, therefore, possible to trace from ozone concentrations its incursion into the troposphere. The inherent stability of the stratosphere inhibits the vertical mixing of aerosols and therefore they have a long residence time, typically of the order of 1–2 years, compared with some 10 days in the troposphere. This fact has important consequences for the climatic effects of volcanic aerosols. The El Chichon explosion of 1982, though of smaller magnitude than the Mount St. Helens eruption of 1980, injected considerably more volcanic aerosol into the stratosphere. The dust cloud, composed mainly of sulphur particles and sulphuric acid droplets, was tracked across the Pacific Ocean for many months following the explosion and it produced significant reductions in solar radiation in Hawaii, over 6000 km from the volcano in Mexico, as has been previously stated. Man-made pollutants will have similar long residence times in the stratosphere and it is in this region, therefore, that their climatic effects may be first observed. Of particular concern has been the input of exhaust gases from high-flying, supersonic aircraft which may initiate photochemical reactions resulting in the reduction of the ozone concentration in the stratosphere and a subsequent increase in the amount of ultraviolet radiation incident on the Earth's surface.

The mesosphere contains $0 \cdot 1\%$ of the mass of the atmosphere and it is less stable than the stratosphere, with temperatures decreasing with height to reach a minimum value at 80 km. The temperatures are sometimes sufficiently low to produce freezing of even the low concentration of water vapour present and to produce noctilucent clouds. These clouds are normally translucent to the direct solar radiation incident upon them but they can sometimes be seen in summer. Above the mesopause is the thermosphere, where the temperature increases steadily with height. In this region, the molecules become very well separated and collisions become so infrequent therefore we can no longer treat the atmosphere as an ideal gas. Furthermore, the thermosphere does not behave as a simple fluid, because electrically charged particles will move differently to neutrally charged particles.

The consideration of the actual processes which produce the observed temperature distribution in the lower 100 km of the atmosphere is now in order. The detailed processes of absorption of solar radiation and long-wave planetary radiation in the atmosphere have already been discussed, and this information may be used to obtain a model of the vertical distribution of radiation and temperature. Consider a horizontal slab of the atmosphere, as depicted in figure 3.9. The net rate of heating (or cooling) of the slab will depend on:

(i) The absorption of solar radiation, ΔE.
(ii) The absorption of long-wave radiation from the atmosphere above, ΔL_{d}, and below the slab, ΔL_{u}.
(iii) The re-emission of long-wave radiation at the slab temperature, L_{e}.

Hence, following the above notation the energy budget of the slab is given by:

$$mC_p \frac{\mathrm{d}T}{\mathrm{d}t} = \Delta E + \Delta L_{\mathrm{u}} + \Delta L_{\mathrm{d}} - 2L_{\mathrm{e}} \qquad (3.1)$$

where m is the mass of the slab and C_p is the specific heat of air at constant pressure.

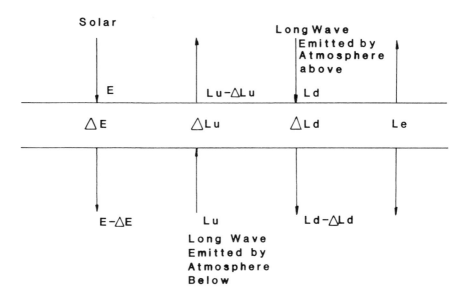

Figure 3.9. Radiation budget of a thin layer of the atmosphere (see text for nomenclature).

In radiative equilibrium, $\mathrm{d}T/\mathrm{d}t = 0$, i.e. the rate of change of temperature with time is zero, and therefore the radiation absorption is balanced by the radiation re-emission. As has been noted earlier, the absorption in a clear, cloud-free atmosphere depends on:

(i) The mass of absorbing gas.
(ii) The path length through the gas.
(iii) The absorption coefficient for the gas for each wavelength.

Ideally, the absorption coefficients would be calculated for each radiation wavelength between $0 \cdot 1$ and $100~\mu\mathrm{m}$, but this is very time consuming, even with high-speed computers, and therefore approximations to the absorption spectra are made. The simplest approximation is to associate a single absorption coefficient with the solar radiation band and another with the long-wave planetary radiation band. Similarly for emission, an effective emissivity, ϵ, can be defined, which depends on the mass of carbon dioxide, water vapour and ozone. Hence, if the vertical distributions of these gases are known, then the atmosphere can be divided up into a number of levels and equations similar to equation 3.1 may be written down for each level. The long-wave emission term, L_e, where $L_e = \epsilon\sigma T^4$, can be adjusted by iteration until it balances the other terms in the equation. Figure 3.10 shows the results of this type of calculation for a cloud-free atmosphere, composed of carbon dioxide, water vapour

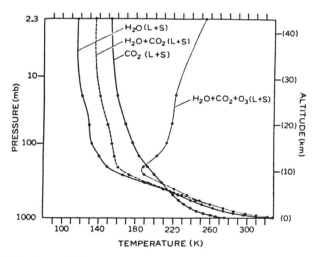

*Figure 3.10. Radiative equilibrium temperature profile for various atmospheric absorbers, in the absence of clouds. The distribution of gaseous absorbers at 35°N in April are used. (L + S) means that the effects of both long-wave radiation and solar radiation are included. (Reproduced from S. Manabe and R. F. Strickler (1964), J.A.S., **21**, 371, figure 6a, with permission of the American Meteorological Society.)*

and ozone. The results of the same calculation for (a) carbon dioxide only, (b) carbon dioxide and water vapour only, and (c) water vapour only are also shown in figure 3.10. It can be seen clearly that, without ozone absorption, the temperature would decrease with height up to 100 km and the stable stratosphere would not exist. This profile is similar to the atmospheric profiles of Venus and Mars, where ozone occurs in insufficient quantities to produce a stratosphere. Another interesting feature is the role of water vapour, which is very important in the absorption in the lower troposphere. In a water-free atmosphere, it is clear that the troposphere and the surface temperature would be considerably cooler than present observations show. It is interesting to note the shielding effect of tropospheric water vapour on the stratospheric temperature. In the absence of water vapour, long-wave radiation is transmitted from the surface into the stratosphere, and hence produces an increase in stratospheric temperature. The carbon dioxide does contribute significantly to the warming in the atmosphere, but, because it is well mixed throughout the atmosphere, it tends to warm the stratosphere as much as the troposphere. It is noted that in the realistic model, as illustrated by figure 3.10, the calculated surface temperature is 330 K, whilst the observed surface temperature is approximately 288 K. The reason for this discrepancy is that the vertical temperature gradient in the lowest 10 km is unstable (i.e. $\partial \theta / \partial z < 0$) for the radiative equilibrium profile. If the model is adjusted to a stable temperature profile ($\partial \theta / \partial z = 0$), then the surface temperature approaches the observed reality and the temperature in the upper troposphere is increased. The assumption inherent in this temperature profile stabilization is that convection will continually re-adjust the vertical temperature profile by transferring heat from the surface to the upper troposphere. Hence, the radiation will always be endeavouring to produce an unstable gradient near the surface but, because of the efficiency of convection in the vertical transfer of heat, the surface will not come into radiative equilibrium. However, very close to the surface, below a height of about 1 m, unstable temperature gradients may be maintained. This occurs when the solar heating of the surface is large and small-scale convection near the surface is not sufficiently vigorous to completely adjust the vertical temperature gradient.

Such model results will be complicated by clouds, aerosols and horizontal variations in the distribution of water vapour and ozone in the atmosphere. It is model calculations such as these, however, which have been used to obtain estimates of the increase in surface temperature as the result of increasing levels of carbon dioxide in the atmosphere.

The fluxes of radiation for the observed global atmosphere are indicated in figure 3.11, which assumes a solar flux of 100 units at the top of the atmosphere, but because of the variability of the absorption and reflection processes, the diagram can act only as a guide to the magnitude

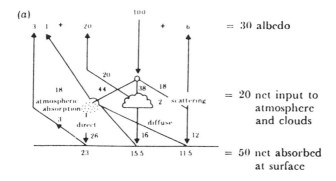

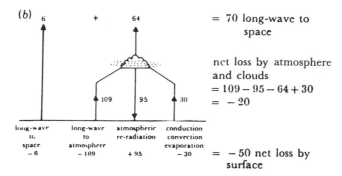

*Figure 3.11. Schematic diagram of the globally and annually averaged components of the short-wave (a) and thermal infra-red (b) radiation streams for the atmosphere. (Reproduced from A. Henderson-Sellers and M. F. Wilson (1983), Phil. Trans. Roy. Soc., **A309**, 286, figure 1, with permission of the Royal Society.)*

of the processes involved. In common with the approach in Sections 3.2 and 3.3, the solar and planetary radiation will be divided into two separate streams. Of the 100 units of solar radiation incident on the top of the atmosphere, about 50 units are absorbed by the Earth's surface, 30 units are reflected back to space, and 20 units are absorbed by the atmosphere. The major reflector of solar radiation is the clouds, whilst the main absorbers are the atmospheric aerosol, water vapour and ozone. It is noted that globally the Earth's surface reflects only 4 units back to space, on account of the low albedo of the oceans. The Earth's surface and atmosphere, to maintain thermal equilibrium, must lose heat at the same rate as it is absorbed from solar radiation. The planetary radiation budget for the surface and atmosphere is rather more complicated than for solar radiation, because of the re-emission of radiation by the atmosphere. For instance 115 units of long-wave radiation are emitted from the Earth's surface, of which 109 units are absorbed by the atmosphere, and the remaining 6 units are lost to space, through the atmospheric window. The atmosphere re-emits 95 units back to the Earth's surface, and therefore the net long-wave flux lost from the surface is only 20 units. Furthermore, the surface gains 50 units by solar radiation, and therefore has a net surplus of 30 units of total radiation (solar + planetary). To maintain the thermal equilibrium of the surface, it is necessary to transfer this surplus heat to the atmosphere by convection and by the evaporation of water. Globally, about 75% of the surplus heat is used to evaporate water, though this fraction will vary between land and ocean. Over land about 50% of the energy is used for evapora-

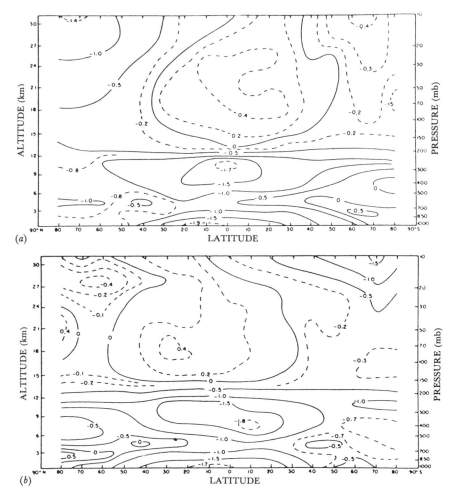

Figure 3.12. (a) Mean total radiative heating (K/day) for December–February. (b) Mean total radiative heating (K/day) for June–August. (Reproduced from T. G. Dopplick (1972), J.A.S., 1291 and 1292, figures 21 and 23, by permission of the American Meteorological Society.)

tion, whilst over the ocean about 90% of the available energy is used. The atmosphere absorbs 109 units of long-wave radiation emitted from the surface, but re-emits 95 back to the surface and 64 to space. Therefore, the atmosphere has a net deficit of long-wave radiation of 50 units. Furthermore, solar absorption by the atmosphere accounts for only 20 units, and therefore there is a net total radiation deficit of 30 units for the atmosphere. The global atmosphere will always be continually cooling by radiation processes, at the rate of 30% of the incident solar radiation kg^{-1} or $\sim 100 \, W \, m^{-2}$. The mass of the atmosphere per unit area is $\sim 10^4 \, kg \, m^{-2}$, and the specific heat at constant pressure is $1000 \, J \, kg^{-1} \, K^{-1}$, and therefore the cooling rate is $10^2/(10^3 \times 10^4) \, K \, s^{-1}$ or $1 \, K \, day^{-1}$. Hence, the atmosphere will continually cool by radiation at $\sim 1 \, K \, day^{-1}$. Therefore to maintain the thermal equilibrium of the atmosphere it is necessary to transfer heat from the Earth's surface by convection and by the release of latent heat of condensation.

Figure 3.12 shows the net heating rates observed for the atmosphere. The continual convection of heat, both horizontally and vertically, implies that the atmosphere will never be in radiative equilibrium. It is noted that, in general, the tropical stratosphere is always heating up, mainly as the result of ozone and carbon dioxide absorption, whilst the troposphere is cooling at rates between $0 \cdot 5$ and $1 \cdot 8 \, K \, day^{-1}$, except over the poles where some heating is observed in the summer hemisphere. The largest cooling rates occur over the surface and at the top of the equatorial troposphere, where high temperatures and high humidities result in a large loss of long-wave radiation. To maintain this cooling rate, heat must be continually convected from the surface into the atmosphere.

3.5. *The absorption of solar radiation in the ocean*

The ocean absorbs the largest fraction of the solar radiation incident on the Earth's surface, because of its great area and its low albedo. The manner in which this radiation is absorbed is of importance in understanding both the heating distribution in the ocean and the stability of the ocean. Furthermore, the biological activity of the ocean is primarily dependent on the direct absorption of the incident solar radiation by microscopic phytoplankton. The rate of phytoplankton growth, or primary production, is a major control on the development of food chains which ultimately determine the productivity of the fisheries and the population of the larger sea mammals at the ends of these chains.

It can be shown that attenuation of solar radiation by pure sea water can be represented by an exponential law of the following form:

$$E_\lambda(z) = E_{\lambda 0} \exp(-k_\lambda z) \qquad (3.2)$$

where k_λ is the attenuation coefficient at wavelength λ and $E_\lambda(z)$ is the solar radiation intercepted at a depth z below the sea surface. $E_{\lambda 0}$ represents the solar radiation incident on the sea surface, at a wavelength λ. The variation of k_λ with wavelength is depicted in figure 3.13(a) for the solar spectrum. The main feature is that the coefficient varies by a factor of 10^7, with a minimum value at $0 \cdot 46\ \mu$m, in the blue region of the visible spectrum, and maximum values in the ultraviolet and near infra-red regions in the spectrum. The inverse of the absorption coefficient, $1/k_\lambda$, gives a measure of the depth of penetration of the radiation, $z_p(\lambda)$. Substituting in equation 3.2:

$$E_\lambda(z_p) = E_{\lambda 0}\,e^{-1} \quad \text{or} \quad 0 \cdot 37 E_{\lambda 0}$$

Thus the penetration depth, $z_p(\lambda)$, is the depth to which 37% of the incident radiation penetrates. At $0 \cdot 46\ \mu$m, this depth is approximately 55 m, whilst at $0 \cdot 3\ \mu$m, in the ultraviolet, it is 10 cm and at 1 μm, in the infra-red, it is 1 cm. The strong absorption of ultraviolet radiation by sea water was important in shielding early organisms from lethal doses of ultraviolet radiation at a time when the ozone layer in the atmosphere was non-existent, as previously discussed in Section 2.1. Less than 1% of the most penetrating blue radiation will reach 500 m and even the human eye, which has a truly remarkable ability to see objects at radiation levels as low as $50 \times 10^{-6}\ \mathrm{W\,m^{-2}}$, would not be able to detect any solar radiation below a depth of 800 m. For these reasons, the majority of the ocean's depths are dark and do not benefit from solar radiation absorption.

The absorption of the solar spectrum in pure sea water, depicted in figure 3.13(b) shows that the majority of infra-red radiation is absorbed in the first 10 cm of the ocean and that all of it is absorbed within 1 m of the surface. Thus, over 55% of the incident solar radiation is absorbed in the first metre of sea water, but once the infra-red and ultraviolet components have been eliminated, the rate of absorption with depth decreases, progressively removing the red and orange components of the solar spectrum and leaving only the most penetrating blue component. The sea water, therefore, acts as a monochrometer, removing all colours except blue below a depth of 10 m. The blue colour of deep sea water is predominantly the result of the absorption process, rather than Rayleigh scattering from sea-water molecules. A person looking downwards into the ocean sees light which has been back-scattered from successive layers of water. Since all of the other colours except blue have been absorbed very rapidly, a large proportion of the back-scattered light comes from layers where only blue light has penetrated. The blue colour of the ocean surface is also enhanced by the reflection of 'sky light'.

Thus far, only the absorption of solar radiation by pure sea water has

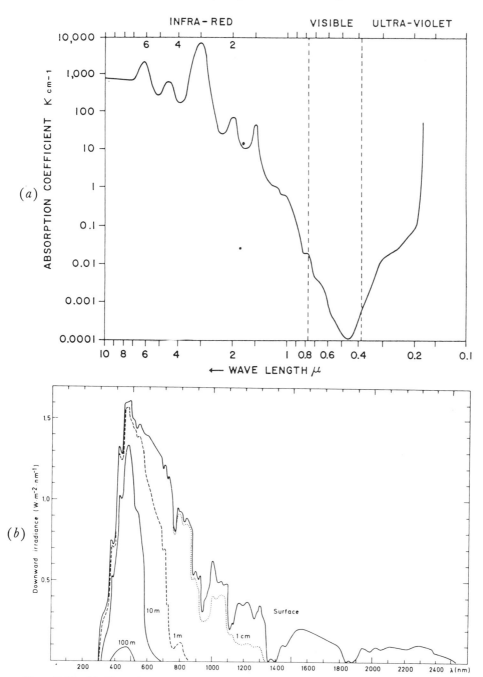

Figure 3.13. *(a) Absorption of radiation by sea water. (b) The complete spectrum of downward solar irradiance in the sea between the surface and 100 m. (Reproduced from* Marine Optics, *by N. G. Jerlov, p.140, figure 7.4, with permission of Elsevier.)*

been considered. Only in regions of very low biological productivity situated at considerable distance from estuarine environments does the pure sea water absorption approach reality. The centres of the large sub-tropical gyres, such as the Sargasso Sea, are such regions of very clear ocean water. In regions of significant primary productivity, the colour of the sea water changes from dark blue to blue-green. The minimum absorption no longer occurs at $0 \cdot 46\,\mu m$, but it is shifted towards the green and occurs at approximately $0 \cdot 48\,\mu m$. The shift is primarily the result of the presence of chlorophyll, which absorbs strongly in the purple and blue, and less significantly in the red part of the spectrum. In addition to the colour change, the presence of chlorophyll also reduces the overall penetration depth of solar radiation. In very productive waters, such as the equatorial upwelling region, the penetration depth for blue light at $0 \cdot 46\,\mu m$ is ~ 10 m, compared with a penetration depth in pure sea water of 55 m. However, the most significant effects are found in coastal waters over the continental shelves. Here, strong tidal mixing and large inputs of suspended solids from rivers increase the attenuation coefficient dramatically and penetration depths of ~ 1 m for blue light are observed. In addition, chlorophyll pigment and other biologically derived pigments produce variations in the perceived colour of sea water. For example, a yellow pigment called gelbstoft, which is prevalent in European coastal waters, has a very strong absorption in the blue part of the spectrum and thus the most penetrating wavelengths occur around $0 \cdot 55\,\mu m$, in the yellow part of the visible spectrum. The presence of large quantities of suspended mineral sediment in some coastal waters tend to impart a grey colour because of scattering by the mineral particles. Occasionally, a red pigment, produced by a particular species of phytoplankton, can cause a phenomenon known as 'the red tide'. This pigment can be remarkably toxic to many other organisms and it is therefore usually associated with a temporary absence of fish and other forms of marine life. Red tides are observed, for example, in Southampton Water, usually after a period of warm weather.

Table 3.1 shows the typical penetration depths of solar radiation for different types of water masses. The most important point to notice is that, because of the presence of biologically derived pigments and mineral and calcareous particles in the ocean, the average penetration depth is much less than that of pure sea water. All of the incident solar radiation is effectively absorbed within the uppermost 50 m of the ocean. Thus, the ocean is warmed from above and the formation of a warm surface layer enhances the stability of the ocean, in the same way that the stability of the stratosphere is enhanced by ozone absorption, and in marked constrast to the troposphere, where radiative heating of the surface maintains instability and convection.

Now, consider the response of the temperature in the upper part of

Table 3.1. Penetration depth (metres) of solar radiation $(0 \cdot 3 - 2 \cdot 4 \ \mu m)$ in typical ocean and coastal water masses.

Water type	Example of water type	Depth for attenuation of 90% of incident solar radiation	Depth for attenuation of 99% of incident solar radiation
Ocean I	Sargasso Sea	33	85
Ocean II	North-eastern Atlantic	14	45
Ocean III	Equatorial East Pacific (upwelling)	8	25
Coastal 1	Continental shelf-break	7	20
Coastal 9	Baltic Sea	2	5

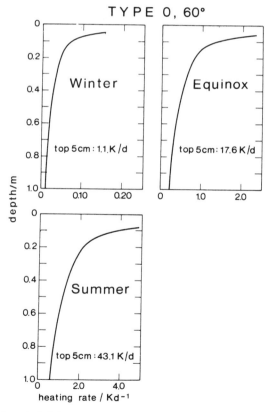

*Figure 3.14. The theoretical daily temperature rise, in the absence of mixing, in the top metre of the ocean for clear sea water in winter, spring and summer at 60° latitude. (Reproduced from J. D. Woods et al. (1984), Q. J. Roy. Met. Soc., **110**(465), 640, figure 3a, with permission of the Royal Meteorological Society.)*

the ocean to the absorption of solar radiation. Figure 3.14 shows the rate of heating calculated in clean ocean water for different seasons at 60° latitude. The heating rates are high, approaching 40 K day^{-1} in the top 5 cm of the ocean during summer because of the rapid absorption of infra-red radiation. Obviously the ocean surface responds relatively quickly to solar heating, but these rates are not generally observed in the ocean for two main reasons. First, the ocean surface will tend to cool as a result of a net loss of long-wave radiation to the atmosphere and to space, by evaporation and by turbulent heat convection. All of these processes will cool the surface skin of the ocean, i.e. the uppermost millimetre of the ocean surface and long-wave radiation of 10 μm wavelength will be emitted from a surface layer approximately 50 μm in thickness. Thus, an unstable temperature profile will be produced and mixing of the surface skin with deeper waters will occur. Secondly, turbulence in the atmosphere will produce surface waves and turbulence in the upper ocean, which will both act to break the surface skin periodically. The warm surface layer under moderate winds of about 5 m s^{-1} will therefore be mixed with cooler sub-surface waters, to depths of over 10 m. Hence, only in regions of very low winds under clear sky conditions will large changes in surface temperature, i.e. changes greater than 1 K day^{-1}, occur. Such regions have been observed by satellites using infra-red detectors, and in one reported case the surface temperature was observed to change by 4 K day^{-1}. Generally, however, the diurnal changes in temperature are observed to be less than 1 K day^{-1} and are often about 0·1 K day^{-1}.

4
Water in the atmosphere

4.1. Introduction

The atmosphere is the smallest reservoir of water on the planet Earth, contributing only $0 \cdot 001\%$ of the total mass of water present. In comparison, the ocean contains 97%, the ice caps of Antarctica and Greenland contain 2.4% and the freshwater reservoirs contribute the remaining percentage. However, the effect of the water present in the atmosphere is by no means small. It has already been shown that water vapour in the atmosphere is an important absorber of both solar and long-wave planetary radiation, and that, upon condensation, liquid water drops can both reflect solar radiation back into space and intercept long-wave radiation from the surface and the atmosphere. Water vapour is a gas with a molecular weight of 18 a.m.u. compared with a mean molecular weight of 29 a.m.u. for dry air, and hence a mixture of water vapour and dry air will have a lower density than an equivalent volume of dry air, assuming that temperature and pressure remain constant. The equation of state for moist air therefore requires another variable, namely the mass of water vapour, to determine the relationship between pressure, temperature and density.

When water vapour condenses there is a large release of latent heat. For each kilogram of condensed water produced, $2 \cdot 4$ MJ of energy are released. This input of energy into the atmosphere is equivalent to slightly less than one-quarter of the solar flux incident on the Earth or approximately 80 W m^{-2} on average, and it is, therefore, a major source of energy for driving the circulation of the atmosphere. For example, the trade-wind circulations of both hemispheres are enhanced by the release of latent heat in the inter-tropical convergence zones (ITCZ), which straddle the thermal equator. The latent heat of condensation of water vapour may be released many thousands of kilometres away from where the evaporation of the water molecule took place and the wind circulation in the lower troposphere is important in determining the location of the energy release. Globally, the precipitation rate must balance the evaporation rate, but relatively small variations in the wind circulation can drastically influence the spatial distribution of precipitation and produce devastating effects on agriculturally based economies.

4.2. The moist atmosphere

Figure 4.1 shows the phase diagram for water within the range of atmospheric temperatures. The curve *AB* is the saturated vapour pressure (SVP) over a plane ice surface and the curve *BC* corresponds to the SVP over a plane liquid water surface. The point *B* represents the point where water vapour, liquid water and ice can co-exist in equilibrium. It is known as the triple point. The main feature to notice is the range of vapour pressure which can exist in the atmosphere. At temperatures of − 50°C, typical of winter-time in the polar regions, the SVP over ice is only 4 Pa, whilst at 30°C, typical of the equatorial region, the SVP approaches 4300 Pa. Between 0 and 30°C, the SVP approximately doubles for each 10°C rise in temperature. In the atmosphere, liquid water can exist in a supercooled state. The SVP over a supercooled water surface is higher than the SVP over an ice surface, thus, a vapour in equilibrium with a super-cooled

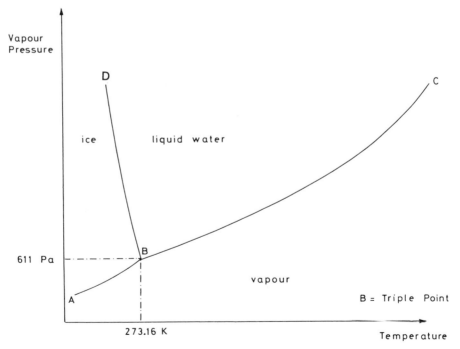

Figure 4.1. Phase diagram for water in the vapour, liquid and solid phases showing the triple point.

water surface would be supersaturated with respect to an ice surface and water vapour would condense preferentially onto the ice surface (figure 4.9). This is an important process in the formation of precipitation in cold clouds, which will be discussed in Section 4.5.

The equation of state for moist air gives the mean molecular weight, $\bar{m}$, of a mixture of dry air of mass M_d and of water vapour of mass M_v in a given volume as:

$$\frac{1}{\bar{m}} = \frac{M_d}{m_d(M_d + M_v)} \left[1 + \frac{M_v/M_d}{m_v/m_d} \right] \qquad (4.1)$$

where m_d and m_v are the molecular weights of dry air and water vapour, respectively.

The concentration of water in the atmosphere is defined by a mixing ratio, q, where

$$q = M_v/M_d \qquad (4.2)$$

The mixing ratio is actually dimensionless, but meteorologists often express it in terms of grams of water vapour per kilogram of dry air. Since, for a given volume, masses are proportional to densities, the above equation may be rewritten as:

$$q = \varrho_v/\varrho_d \qquad (4.3)$$

where ϱ_v and ϱ_d are the densities of water vapour and dry air, respectively. The ideal gas equations for dry air and water vapour yield the following:

$$\varrho_v = \frac{e}{(R^*/m_v)T} \qquad (4.4)$$

and

$$\varrho_d = \frac{p - e}{(R^*/m_d)T} \qquad (4.5)$$

where p is the total pressure of the dry air and vapour and e is the vapour pressure of water. Water vapour is not in general an ideal gas, but at low concentrations, such as found in the atmosphere, the approximation is a valid one.

Substituting equations 4.4 and 4.5 into 4.3 yields:

$$q = \frac{e}{p - e} \left(\frac{m_v}{m_d} \right) \qquad (4.6)$$

An appropriate expression for q can be obtained by noting that the total pressure of the dry air and the water vapour is usually two orders of magnitude larger than the vapour pressure of water, and hence

$$q \simeq \frac{e}{p} \left(\frac{m_v}{m_d} \right) \tag{4.7}$$

Substituting for M_v/M_d in equation 4.1 gives:

$$\frac{1}{\overline{m}} = \frac{1}{m_d} \left[\frac{1 + q(m_d/m_v)}{1 + q} \right] \tag{4.8}$$

Substituting equation 4.8 into the equation for a mixture of ideal gases, the ideal gas equation for a moist atmosphere is obtained, namely:

$$p\alpha = \frac{R^*}{m_d} \left[\frac{1 + q(m_d/m_v)}{1 + q} \right] T \tag{4.9}$$

This equation is similar in form to the ideal gas equation for dry air except for the addition of the factor involving the mixing ratio, q. An example of the use of this equation is now appropriate. If the vapour pressure of water in the atmosphere is 1000 Pa and the total pressure is 10^5 Pa then q is found to be $6 \cdot 22 \times 10^{-3}$ or $6 \cdot 2$ g of water vapour per kilogram of dry air, and hence

$$p\alpha = \frac{R^*}{m_d} [1 + 3 \cdot 78 \times 10^{-3}] T$$

At this stage it is useful to define a virtual temperature, T_v, where

$$T_v = T \left[\frac{1 + q(m_d/m_v)}{1 + q} \right] \tag{4.10}$$

This virtual temperature is always higher than the actual temperature, but in practice, the difference is less than 5 K. In middle latitudes, the difference is typically 1 K. In physical terms, the virtual temperature is the temperature a volume of dry air would have if it had the same density as an equivalent volume of moist air, and because water vapour is a lighter gas, this temperature will always be larger than the actual temperature.

Another useful parameter is the relative humidity, r, given by:

$$r = \frac{q(T)}{q_s(T)} 100\%$$

where q is the actual mixing ratio of the sample of air and q_s is the saturation mixing ratio at the same temperature. It gives a useful measure of the humidity of moist air mass relative to its saturation value and hence 100% relative humidity corresponds to a saturated air mass.

4.3. *Measurement and observations of water vapour*

Direct measurements of the vapour pressure and the mixing ratio, q, though feasible in a laboratory, are not practicable for the routine measurements in the atmosphere. The most direct method is to pass air over a flat silver plate which is cooled and the temperature at which a dew first forms is noted, i.e. the dew-point temperature, T_d, is measured. The water-vapour content of the air is unaltered during the cooling, as shown in figure 4.2, and hence the saturated mixing ratio at the dew-point temperature is equal to the mixing ratio of the air at its original temperature, T. The saturated mixing ratio is a known function of dew-point temperature and it can, therefore, easily be obtained from tables. It is noted that the cooling process must take place at constant atmospheric pressure to maintain a constant mixing ratio, as shown in equation 4.6. At lower temperatures, frost

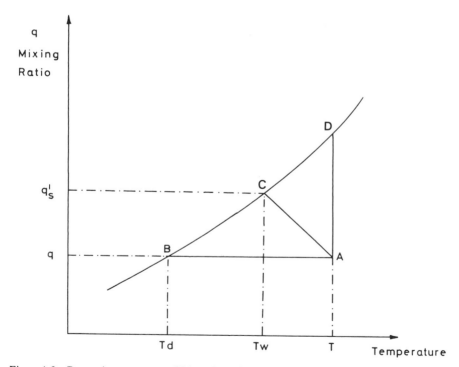

Figure 4.2. Dew-point temperature (T_d) *and wet-bulb temperature* (T_w), *for a parcel of moist air* (A)*at temperature* T *and mixing ratio* q. *BCD is the saturation mixing ratio,* q_s. *The region above BCD is supersaturated whilst the region below is unsaturated. Dew point temperature,* T_d, *is obtained by cooling without changing the mixing ratio* (AB). *Wet bulb temperature,* T_w, *is obtained by evaporation of water into air until saturated* (AC).

is more likely to form rather then dew and in this case the saturated mixing ratio with respect to ice must be used instead of that with respect to water.

A more practical method for surface measurements of humidity is that using the aspirated psychrometer. In this technique, air is drawn past two thermometers, one of which has its bulb covered with a muslin cloth, which is kept moist with distilled water. The wet-bulb temperature always lies between the dry-bulb temperature and the dew-point temperature, except at saturation, when all three temperatures coincide, as shown in figure 4.2. The wet bulb evaporates water into the surrounding air, which:

(i) Decreases the temperature of the wet bulb due to the latent energy input required to produce evaporation.
(ii) Increases the mixing ratio of the surrounding air.

Hence the saturated air surrounding the wet bulb can be represented by a point C in figure 4.2, and the difference between the saturation mixing ratio at the wet-bulb temperature (q'_s) and the mixing ratio of the environmental air (q) can be shown to be proportional to the difference between the wet- and dry-bulb temperatures, i.e.

$$T - T_w = A[q'_s - q]$$

where A is a constant which depends on the ventilation rate of the instrument. q'_s can be obtained from tables and hence q is obtained.

Other techniques include the measurement of the change in length of hair or skin with changing humidity. A human hair will expand by 2% in changing from an absolutely dry to a saturated environment and this property can be used to measure relative humidity where cheap and reasonably reliable instrumentation is required. This technique is used in some types of radiosonde, but it cannot be used to measure low humidities and the hair component tends to exhibit strong hysteresis after becoming saturated in clouds.

The distribution of water vapour in the atmosphere is shown in figure 4.3. The mixing ratio reaches a maximum in the equatorial zone and rapidly decreases with both altitude and latitude. This is a general reflection of the variation of the saturated mixing ratio with temperature. The majority of the water vapour occurs below 500 mb, although the small amounts present in the upper troposphere and the stratosphere have already been shown to be important for the radiation balance in these regions. If all the water vapour were to be condensed out of the atmosphere, it would result in an average global precipitation of $2 \cdot 5$ cm, ranging from 4 cm in the equatorial regions to less than $0 \cdot 5$ cm over the poles. This can be compared with the observed global precipitation of 100 cm year^{-1}. The ratio of the water content of the atmosphere to the observed precipitation rate

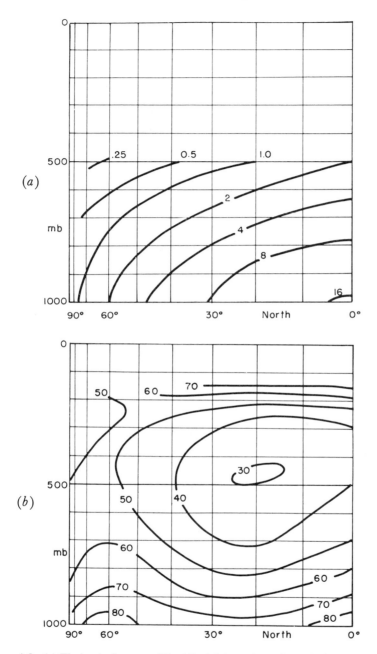

Figure 4.3. (a) The longitude-averaged humidity (q) in northern winter (October–March). Values are in grams/kilogram of dry air. (b) The longitude-averaged relative humidity (q/q_s) in northern winter. Values are in per cent. (Reproduced from The Nature and Theory of the General Circulation of the Atmosphere, *by E. N. Lorenz, World Meteorological Organization 218 TP 115, p. 42 and 43, figures 14 and 16, by permission of WMO.)*

gives the time scale for the replacement of the atmospheric water vapour by evaporation from the surface. Thus, the average lifetime of a water molecule from evaporation at the surface to precipitation is approximately 9 days and this is sufficient for the water molecule to be carried around the globe by the tropospheric wind system before it is precipitated. The relative humidity distribution shown in figure 4.3 gives an indication of the efficiency of the water-holding capacity of the atmosphere. In the equatorial zone, between 0 and $10°$, and at latitudes greater than $40°$, the atmosphere is close to saturation. However, in the sub-tropical zone the relative humidity is very low, attaining values of only 30%, as a result of the descent of air from the upper troposphere, where it has a very low water content. This air will warm by adiabatic compression as it moves downwards through the troposphere but will retain its low water content. Hence, the relative humidity will decrease as one moves from the upper to the lower troposphere, although close to the surface, evaporation into the air mass will act to increase the relative humidity.

4.4. Stability in a moist atmosphere

It has already been shown that water vapour will change the thermodynamic properties of a sample of air by:

(i) Decreasing the mean molecular weight of the moist sample in comparison with a sample of dry air.
(ii) The release of the latent heat of condensation when precipitation from the moist sample occurs.

The change in the first property is directly proportional to the mass of water vapour in the given sample of air, whilst the second property is proportional to the mass of water vapour which condenses to form liquid water. In order to obtain a qualitative feel for the relative effects of the two processes, figure 4.4 illustrates the temperature history of a parcel of moist air, typical of the middle latitudes, as it is lifted through the atmosphere. At the surface, it is assumed that the air has a water-vapour content of $3 \cdot 5$ g per kilogram of dry air, which corresponds to a virtual temperature 1 K higher than the actual temperature. From the surface to a height of 1 km, the parcel of air cools at the dry adiabatic lapse rate† of $9 \cdot 8$ K km^{-1}. At the point at which the condensation level is reached, release of latent heat slows down the rate of cooling of the parcel. The amount of latent heat released between 1 and 3 km depends solely on the difference between the saturation mixing ratios of the air at the two levels, which is approximately

---

†A moist unsaturated sample of air will cool at the dry adiabatic lapse rate, to a good approximation.

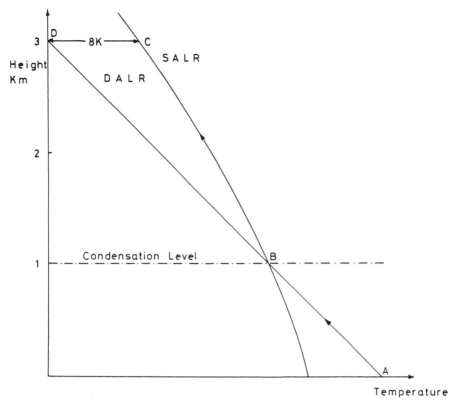

Figure 4.4. Typical ascent of a moist air parcel (ABC) *and a dry air parcel* (ABD). *See text for details.*

7 g per kilogram of dry air in this case. This is equivalent to the release of 17,500 J for each kilogram of dry air. The temperature difference between a moist parcel of air at 3 km and a dry parcel of air at the same height is approximately 8 K which is due solely to the release of the latent heat of condensation. It is noted that the difference between the actual and the virtual temperature is generally smaller, though not negligible, compared with the difference between dry air parcels and moist air parcels after condensation.

Now consider the change in temperature with height of a parcel of air at the level of condensation. If the parcel rises an infinitesimally small distance, dz, then the difference in the saturation mixing ratio, brought about by adiabatic cooling, will be $-dq_s$. The negative sign denotes a decrease in the saturation mixing ratio with increasing height. The latent heat released for each kilogram of dry air is given by $-Lq_s$, where L is the latent heat of condensation, and this heat is, therefore, an additional source of energy, equivalent to the term dQ in equation 2.5. Thus:

$$- L \, dq_s = C_p \, dT - \alpha \, dp \qquad (4.11)$$

and from the hydrostatic equation:

$$\alpha \, dp = - g \, dz \qquad (4.12)$$

hence, by substitution in equation 4.11:

$$- L \, dq_s = C_p \, dT + g \, dz \qquad (4.13)$$

Rearranging and dividing through by dz yields

$$\frac{dT}{dz} = - \frac{L \, dq_s}{C_p \, dz} - \frac{g}{C_p}$$

Since $dq_s/dz = (dq_s/dT)(dT/dz)$

$$\frac{dT}{dz} = \frac{-g}{C_p} \left[\frac{1}{1 + (L/C_p)(dq_s/dT)} \right] = \Gamma_s$$

where Γ_s is known as the saturated adiabatic lapse rate (SALR).

Figure 4.9 shows that dq_s/dT is always positive and therefore the SALR will always be less than the dry adiabatic lapse rate, given by $\Gamma_d = - g/C_p$. At a temperature of 20°C and at pressure of 10^5 Pa, the SALR is approximately $0 \cdot 41 \Gamma_d$ or $4 \, \mathrm{K \, km^{-1}}$, whilst at 10°C it is $0 \cdot 59 \Gamma_d$, or $5 \cdot 8 \, \mathrm{K \, km^{-1}}$. At high levels in the troposphere, the saturation mixing ratio will become very small and the SALR will approach the dry adiabatic lapse rate asymptotically.

The addition of latent heat of condensation also increases the potential temperature of the air, and hence in any region of the atmosphere where condensation occurs, the potential temperature will no longer be conserved. The stability of the atmosphere will therefore be significantly changed by the condensation process. Consider a parcel of dry air displaced from its equilibrium position in a stable environment where the temperature gradient, $d\theta/dz$, is positive. As discussed in Section 2.6, the parcel will be subjected to a negative buoyancy force as it rises, and this force will act to return it to the equilibrium position. Now consider a parcel of air which is saturated at the equilibrium position. On rising a small distance δz, it will condense some of its water vapour and this will warm the parcel by an amount $\delta \theta_1$. If $\delta \theta_1 > \delta \theta$ (where $\delta \theta$ is the temperature change of the environmental air over a small distance δz), then the moist parcel will be warmer than its environment (path AB in figure 4.5), and it will accelerate upwards away from its equilibrium position. The result of the condensa-

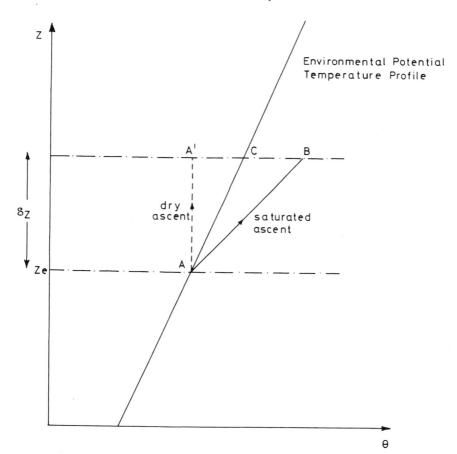

Figure 4.5. Conditional instability of a moist atmosphere. Unsaturated ascent (AA′) is stable but saturated ascent (AB) is unstable.

tion process is therefore to make the previously stable environment profile unstable. Clearly, if the warming is insufficient, i.e. $\delta\theta_1 < \delta\theta$ the profile will still be stable. This form of instability is known as 'conditional instability' and is produced by the condensation process. However, it is not possible to determine the conditions for instability from the potential temperature and an equivalent potential temperature, θ_e, must be defined:

$$\theta_e = \theta \exp\left(\frac{Lq_s}{C_p T}\right)$$

This equivalent potential temperature is conserved in saturated conditions and it is defined as the potential temperature a parcel of air would have if all the water vapour in the parcel were to be condensed, and all the latent heat of condensation converted into internal energy. It has the property of remaining constant during a saturated adiabatic process. Thus, the stability condition for saturated air is determined from the vertical profile of the equivalent potential temperature rather than from the potential temperature.

If, in the case considered above, $d\theta_e/dz < 0$, then the vertical profile will be unstable for saturated air. Parcels of air would have positive buoyancy when they rose and they would accelerate away from their initial surroundings. A profile with $d\theta_e/dz > 0$ would be stable for both dry and saturated air. It is therefore possible to compile a simple stability table:

	$d\theta_e/dz < 0$	$d\theta_e/dz > 0$
$d\theta/dz < 0$	Always unstable	Not possible
$d\theta/dz > 0$	Conditionally unstable	Always stable

It is noted that the stability of an environmental temperature profile can be determined either from the equivalent potential and the potential temperature profiles or by comparison of dry adiabatic and saturated adiabatic lapse rates.

An additional form of moist instability is termed the potential instability and this occurs when the profile has a negative equivalent potential temperature gradient but the air is not saturated. In this case, the lifting of the air mass is required to cool the air to its condensation point and to release the instability. This type of instability is often associated with the advection of an upper level, dry air mass over a very moist, low level air mass. The advection produces a large negative equivalent potential temperature gradient. Usually uplift by mountains or uplift ahead of a depression may be sufficient to release the instability, which is generally associated with vigorous convection and often produces severe thunderstorms with hail.

It is noted that a vertical gradient of humidity may induce instability in unsaturated air. Consider a neutral temperature profile onto which is superimposed a humidity gradient. In this case the virtual temperature would be higher nearer to the surface than above and parcels of air would have positive buoyancy if they were displaced upwards, and would, therefore, accelerate upwards. This type of instability is probably most important in the surface boundary layer over the tropical ocean where the virtual temperature can be several degrees higher than the actual temperature. It is also possible to introduce instability by the evaporation of precipitation in unsaturated air. Precipitation falling from a cloud may evaporate and cool the air beneath to the wet-bulb temperature. The air will become negatively buoyant and accelerate downwards. This process is responsible for the sudden drop in temperature experienced in the downdraughts of thunderstorms and for the sudden gusts of wind, up to 20 m s^{-1}, felt at the surface. Re-evaporation of precipitation is particularly common in the tropics, where cloud bases are generally higher than those found in the middle latitudes.

4.5. *Processes of precipitation and evaporation: the formation of clouds*

In the previous discussion it was implicitly assumed that condensation would occur when the relative humidity reached 100%, i.e. when the vapour pressure was equal to the saturation vapour pressure. However, the saturation vapour pressure described in Section 4.2 was defined with respect to a plane liquid water surface and therefore the definition is only strictly applicable to the formation of dew. In a sample of pure moist air, condensation will not occur at 100% relative humidity but will only commence spontaneously at a relative humidity of 800%, equivalent to a supersaturation of 700%. The reason for this phenomenon is that the saturated vapour pressure over a curved surface is higher than over a plane surface. The relationship between the saturated vapour pressure over a curved surface e_r and that over a plane water surface e_∞ is given by:

$$\ln \frac{e_r}{e_\infty} = \left(\frac{2\gamma m_w}{\varrho_w R^* T} \right) \frac{1}{r} \tag{4.14}$$

where γ, m_w and ϱ_w are the surface tension, molecular weight of water and density of water, respectively, r is the radius of curvature and R^* is the universal gas constant. At a temperature of 273 K, the term in the bracket is $1 \cdot 19 \times 10^{-9}$ m.

If the value $e_r/e_\infty = 8$ is substituted in equation 4.14, a value of $r \approx 0 \cdot 6 \times 10^{-9}$ m is obtained, and this is the typical size of a small group of water molecules. Therefore, condensation in pure air samples will occur on groups of water molecules randomly brought together by Brownian motion. However, if a sample of air is taken directly from the atmosphere, condensation will occur at much smaller supersaturations, typically less than 1%. This is because of the presence of particles in the atmosphere which provide surfaces for condensation. These particles are known as condensation nuclei. The size of these nuclei, estimated by substitution into equation 4.14, is approximately $0 \cdot 1 \ \mu m$ for a supersaturation of 1%. Basically, there are two types of condensation nuclei. The first type are hygroscopic nuclei and they are usually ionic salts, such as sodium chloride and ammonium sulphate, which dissolve readily when condensation occurs. They have the property that they are able to initiate condensation at relative humidities as low as 80%. The second type of condensation nuclei are the hydrophobic particles which are not soluble in water and which require supersaturation conditions for condensation to occur.

Equation 4.14 has to be modified for hygroscopic nuclei. Figure 4.6 shows the SVP curves for pure water droplets and droplets containing dissolved hygroscopic nuclei. It is noted that the difference between the two SVP curves is most marked for small particles having radii less than $0 \cdot 1$ μm. For larger water droplets, the hygroscopic nuclei will become well diluted and the effect becomes less pronounced. In addition, the radius of curvature also becomes large enough to approximate to a plane surface and for large water droplets, bigger than 1 μm, the difference between the two types of nuclei is no longer important. One interesting aspect of the difference in behaviour between hygroscopic and hydrophobic nuclei can be seen by comparing the growth of two small droplets. In an environment where the relative humidity is less than 100%, only hygroscopic nuclei will present possible condensation sites, but as the water droplet grows the salt becomes more dilute and a higher relative humidity is required for further growth. This process is represented by path *AB* in figure 4.6, and

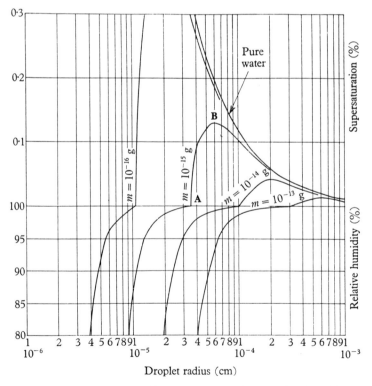

Figure 4.6. The equilibrium relative humidity (or supersaturation) as a function of droplet radius for solution droplets containing the indicated mass of sodium chloride. (Reproduced from Clouds, Rain and Rainmaking, *by B. J. Mason, p. 21, figure 4, with permission of Cambridge University Press.)*

once the droplet has reached the size dictated by the prevailing relative humidity, its growth will cease. Only if there is an increase in the relative humidity of the environmental air, or supersaturation is achieved, will further growth occur. This situation is typical for the development of haze. A maritime haze layer will develop for a few hundred metres above the surface of the ocean, where salt nuclei are readily available because of bubble bursting at the sea surface. Haze over land may form as the result of the introduction of hygroscopic pollutants from industrial processes or by photochemical reactions of the gaseous car exhaust emissions.

In an environment where there are very few hygroscopic nuclei, supersaturation of the air, of the order of 1%, will be required to start the condensation process. However, once a droplet starts to grow its radius will increase, the SVP (with respect to the curved surface) will decrease and the droplet will find itself in a continually saturated environment and will carry on growing. Thus, once a critical supersaturation has been reached, droplets will grow, but as the supersaturation curve approaches 100% relative humidity, the rate of growth will decrease. Water droplets larger than 10 μm in radius grow only very slowly by this process.

The availability of condensation nuclei depends on the rate of supply of suitable particles and on their rate of removal, either by gravitational settling or by 'washout'. Figure 4.7 shows the typical sources and sinks of atmospheric aerosols and the size distributions for the various types of aerosol. The small Aitken nuclei, less than 0·1 μm in radius, tend to be produced by smokes and gases and are plentiful in industrial areas. They tend to be too small to be efficient as condensation nuclei on their own, but coagulation can produce larger nuclei, which are very efficient condensation nuclei. The majority of the aerosol is removed by 'washout', whilst the remainder is removed by dry deposition or gravitational settling. A particle 1 μm in diameter would take about 1 year to fall 1 km, whilst a 10 μm particle would take only 3·5 days and thus gravitational settling is an important process only for the deposition of giant aerosols. For smaller Aitken nuclei, scavenging by precipitation will be the principal removal process.

The number of nuclei reduces from 10^{12} m^{-3} over industrial areas to about 10^{8} m^{-3} over the ocean, where the dominant nuclei are salt particles. These are principally produced by bursting bubbles in foam patches which are produced, in their turn, by breaking waves. The number of particles entering the marine atmosphere is dependent on the area of foam over the ocean and this is determined by the sea state and the surface wind speed. The number of particles can vary from 10^{6} m^{-3} in low wind conditions to 10^{9} m^{-3} in storm force conditions. Many of the larger particles will re-enter the ocean but a small proportion will be taken aloft by atmospheric turbulence. These salt particles can have a very large range and they may be found in clouds many hundreds of kilometres away from

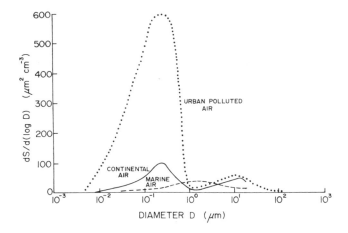

Figure 4.7. Schematic curves of aerosol surface area distributions for urban polluted air, continental air and marine air. Shown below the curves are the principal sources and sinks of atmospheric aerosol and estimates of their mean lifetimes in the troposphere. (Reproduced from N. Slinn and W. George (1975), Atmos. Environ., *9, 763, figure 1, with permission of Pergamon Press.)*

the ocean. As a general rule, there are fewer large nuclei over the ocean than over the continents. The giant aerosols are produced mainly by 'mechanical' wind action at the ocean or land surface under storm conditions. Wind-blown sand, called aeolian, can be transported considerable distances and red Saharan dust is occasionally reported in the British Isles and in the Caribbean, over 3000 km from the Sahara. In addition, large industrial plants produce considerable numbers of giant nuclei.

The typical size distribution of water droplets in clouds may now be considered. It may be obtained by exposing an oil-covered slide from an aircraft and measuring the droplet population with a microscope. A typical distribution is described by three parameters:

(i)   The mean radius.
(ii) The median radius.
(iii) The maximum radius.

Table 4.1 shows the statistics of drop distributions in various cloud types.

Table 4.1. Cloud drop distributions and liquid water content in typical maritime and continental clouds.

Cloud type	Droplet number density (cm^{-3})	Range of radii of droplets (μm)	Most frequent radius of droplet (μm) spectrum	Liquid water content (g m^{-3})
Continental cumulus	420	3–10	6	0·4
Maritime cumulus	75	2–20	11	0·5
Cumulonimbus	72	2–100	5	2·5
Altostratus	450	1–13	4	0·6
Continental stratus	260	1–22	4	—
Maritime stratus	24	2–45	13	0·35

In continental clouds there is a much larger number of droplets than in ocean clouds, because there is an increased population of large condensation nuclei over land, whilst the amount of water vapour available for condensation is determined by large-scale processes, such as uplift, thus ensuring the liquid water contents of maritime and continental clouds are similar, the sizes of the droplets in continental clouds is smaller than in maritime clouds. It is noted from table 4.1 that the maximum droplet size in maritime clouds can be twice as large as those in continental clouds. This observation is of critical importance in the determination of the time scale for the growth of droplets in clouds.

Table 4.1 also shows that the maximum droplet size is related to the type of cloud. Weak, vertical motions in the stratocumulus result in a reduction in the maximum size whilst the strong vertical motions found in cumulonimbus result in large droplets. The water content of cumulonimbus clouds is much larger than in cumulus clouds because of the different surface-to-volume ratio of the two types of cloud. In cumulus cloud, usually 100 m to 1 km thick, this ratio is large and mixing of dry air across the surface of the cloud will significantly modify the water content. In the cumulonimbus the ratio will be smaller because of the size of the cloud, typically 5–10 km thick, which implies that mixing across the surface of the cloud will have less effect in reducing the water content.

A qualitative discussion of the growth of a small water droplet in a supersaturated environment has already been given. This growth is dependent on (*a*) the supersaturation of the environment, (*b*) the size of the drop and is related to the rate of diffusion of water vapour onto the droplet. As the droplet grows, the curvature effect on the SVP decreases and therefore the rate of growth decreases. The growth of a small droplet, 1 μm in diameter,

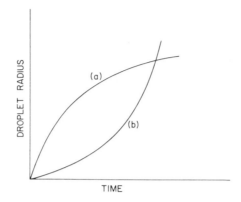

Figure 4.8. Schematic curves of droplet growth by (a) condensation from the vapour phase and (b) coalescence of droplets. (Reproduced from Atmospheric Science, *by J. M. Wallace and P. V. Hobbs, p. 171, figure 4.16, with permission of Academic Press.)*

is depicted in figure 4.8. It takes approximately 30 min for the droplet to reach 10 μm, a further $1\frac{1}{2}$ hours to reach 20 μm and 10 hours to grow to 40 μm. Even a 40 μm droplet is a long way short of the 200 μm of a respectable drizzle droplet. Clearly the diffusion process cannot account for the growth of raindrops in the time scale of a shower cloud, which is typically $\frac{1}{2}$–2 hours. The additional process is the growth of raindrops by collision and coalescence. A droplet may collide with another drop, coalesce and increase its mass. Its vertical velocity will increase and it will sweep out more droplets at a faster rate and it will soon reach a precipitation size. The process is very non-linear in that initial growth will be slow, because of the low terminal velocity of the small droplet, but the growth rate will become larger and larger as the droplet grows. This is shown in figure 4.8.

The rate of increase in mass of a droplet is given by:

$$\frac{\mathrm{d}M}{\mathrm{d}t} = \pi r_1{}^2 (v_1 - v_2) q_L E_c \qquad (4.15)$$

where r_1 is the radius of the droplet, $v_1 - v_2$ is the difference between the terminal velocities of the growing droplet and the surrounding cloud droplets, respectively, q_L is the liquid water content and E_c is the collision efficiency. Except in the initial stage of growth, $v_1 \gg v_2$ and substituting for $M = \frac{4}{3}\pi r_1^3 \varrho_L$ into equation 4.15 yields

$$\frac{\mathrm{d}r_1}{\mathrm{d}t} = \frac{v_1 q_L E_c}{4\varrho_L} \qquad (4.16)$$

The terminal velocity, v_1, for different droplets is shown in table 4.2. A

Table 4.2. Terminal velocities of water droplets in still air at surface atmospheric pressure and temperature of 20°C.

Radius of droplet (μm)	Terminal velocity in still air (cm s^{-1})
1	0·01
5	0·3
25	7·2
50	25·6
250	204
500	403
1000	649
2500	909

10 μm water droplet has a terminal velocity of 10^{-2} m s^{-1}, whilst a drizzle droplet has a terminal velocity of about $\frac{1}{2}$ m s^{-1} and a rain drop has a maximum terminal velocity of 9 m s^{-1}. Thus the rate of growth, which is proportional to the terminal velocity, will accelerate with droplet size. The growth rate is also dependent on the liquid water content of the cloud and the collision frequency. As already noted, cumulus and stratocumulus tend to have a lower water content than the larger cumulonimbus and therefore the precipitation process will be more effective in the latter cloud. The collision frequency increases with droplet size and is low for droplets less than 20 μm in radius. When the cloud droplets are small, they tend to follow streamlines around the collector droplets and do not get captured. For larger droplets, of 70 μm radius or more, turbulent wakes develop behind the droplet, drawing in more droplets than were originally in the volume swept out.

The production of precipitation from a water-droplet cloud is a non-linear process. In a given distribution of cloud droplets, only a few droplets will be bigger than 20 μm and thus be sufficient in size to grow efficiently by collision and coalescence. At first, the process will be slow, but, because of the non-linearity of the process (figure 4.8), the droplet will soon grow quickly to the size of a raindrop. In a maritime cumulus, with droplet concentrations of 100 cm^{-3}, droplets will grow from 25 to 1000 μm within 10 min. Of course only a small proportion at the extreme end of the droplet population will be involved in the precipitation process. The differences between the size distributions in maritime and continental clouds has already been noted, as well as the fact that in many cases r_{max} is less in continental than in maritime clouds. A continental cloud will therefore take longer to produce rain than the corresponding maritime cloud and it has been calculated that a maritime cumulus cloud $1\frac{1}{2}$ km deep will produce rain within 30 min of its formation, whilst a similar continental cloud will take more than 1 hour. This illustrates how dramatic an influence the concentration of condensation nuclei has on the rate of growth of water droplets and hence on the formation of precipitation.

The growth of a shower cloud is dependent on:

  (i) The cloud droplet size distribution.
 (ii) The liquid water content of the cloud.
(iii) The thickness of the cloud.
(iv) The updraft within the cloud, which is essential for the growth of cloud droplets into raindrops.

Small droplets have small terminal velocities and will be maintained in the cloudy environment for longer than larger droplets, provided that they are not ejected through the top of the cloud. As the updraft increases, the size of the final raindrop increases, up to a maximum size of 5 mm and at this stage, the raindrop will break up into smaller droplets. A maritime cumulus will have to be at least 1 km thick to produce light precipitation. However, drizzle may be produced by stratus cloud whose thickness is less than 1 km because the weak vertical motion within such clouds allows the droplets to be maintained in the cloudy air for a sufficient time to reach drizzle drop size.

So far, the growth of precipitation in clouds where the temperatures are above $0°C$ has been discussed. These are known as warm clouds and are found as tropical cumulus clouds and as low-level clouds in middle latitudes. However, below $0°C$, additional physical processes come into play as a result of the introduction of the ice phase. Consider a cloud of liquid water droplets with a mean size of 10 μm. As the droplets are cooled below $0°C$ only a few droplets will freeze and the remainder will be in a supercooled liquid state. As the temperature is reduced still further, the number of ice crystals will increase but only when the temperature reaches $-39°C$ will all the water droplets have frozen. At this temperature spontaneous nucleation will occur, caused by the chance aggregation of water molecules into an ice embryo. This process is analogous to the condensation of water vapour in a cloud chamber in the absence of ions or other particles. However, the observation that ice crystals are formed at temperatures higher than $-39°C$ indicates the existence of ice nuclei. These ice nuclei are different from condensation nuclei in two important respects. First, the most successful ice nuclei tend to have hexagonal molecular structures similar to ice. The most common types of ice nuclei are clay minerals, obtained from the soil, and organic materials, particularly decayed plant matter and derivatives of phytoplankton from the ocean. Secondly, the ice nuclei have a threshold temperature at which they become effective and below which they act as preferential sites for ice nucleation. For clay particles this temperature is $-15\,^{\circ}\mathrm{C}$, whilst organic nuclei may have threshold temperatures as high as $-4\,^{\circ}\mathrm{C}$. Thus, the presence of ice nuclei only has a significant effect below $-4\,^{\circ}\mathrm{C}$ and this explains the observation that clouds between 0 and $-4\,^{\circ}\mathrm{C}$ exist in the supercooled liquid state.

The processes which lead to the growth of ice crystals in a supercooled cloud will now be considered. Figure 4.9 shows the SVP over a plane water and a plane ice surface. It is noted that the SVP over ice is always lower than over liquid water, except at $0\,^{\circ}\mathrm{C}$, and that water vapour in equilibrium with a water surface will therefore always be supersaturated with respect to an ice surface below $0\,^{\circ}\mathrm{C}$. The difference in SVP with respect to ice and water reaches a maximum at $-12\,^{\circ}\mathrm{C}$. Here, in a saturated environment with respect to supercooled liquid water, the vapour would have a supersaturation of 10% with respect to ice. In a mixed cloud, consisting of supercooled liquid water droplets and ice crystals, the ice crystals will grow at the expense of the water droplets.

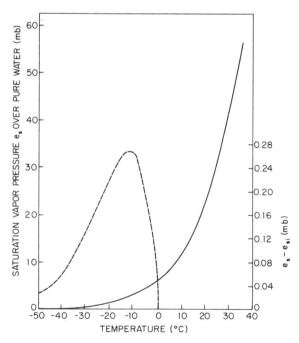

Figure 4.9. Variations with temperature of the saturation vapour pressure, e_s, over a plane surface of pure water (————) and the difference between e_s and the saturation vapour pressure over a plane surface of ice e_{si} (– – –). (Reproduced from Atmospheric Science, *by J. M. Wallace and P. V. Hobbs, p. 73, figure 2.7, with permission of Academic Press.)*

This process is known as the Bergeron process, after the Norwegian cloud physicist who first noted its existence. It is most effective in clouds where the temperature is between -10 and $-30°C$ (i.e. where the difference in SVP is greatest). Between 0 and $-10°C$ there are too few ice nuclei present for the mechanism of ice crystal formation to be effective, whilst below $-30°C$ there are too many ice crystals competing for too few liquid water droplets for the process to be significant. The growth rates of the ice crystals in mixed clouds are much greater than the growth rates of water droplets in warm clouds because of the higher supersaturations occurring in mixed clouds compared with those in warm clouds, typically 10% supersaturation compared with less than 1% supersaturation. For example, an ice crystal can grow to 1000 μm in diameter within 30 min by the Bergeron process.

An ice crystal can grow by two additional processes other than the Bergeron process. First, it can grow by aggregation of ice crystals, because supercooled water films attached to an ice crystal can act as binding for other ice crystals, as a result of surface tension between the water and the ice. The thin film of supercooled water then freezes and the crystal grows. The process is most effective between 0 and $-10°C$, where there are a large number of supercooled water droplets and freezing does not occur immediately on impact. A snowflake may grow from 1 mm to 1 cm in about 30 min by aggregation and the largest snowflakes occur at temperatures close to $0°C$. At lower temperatures, supercooled water droplets may freeze on impact with ice crystals to produce a rimed snowflake. This process is known as accretion, and it is responsible for the growth of ice crystals between -10 and $-30°C$. It is most effective in clouds such as cumulonimbus where the liquid water content is high. In contrast, the Bergeron process is most effective in small cumulus and stratocumulus clouds, where the water content is low. Accretion and the riming of ice particles can produce soft hail and ice pellets and, more exceptionally, hailstones. The accretion process has many similarities with the collision–coalescence mechanism, in that it is very non-linear, due to the increase in terminal velocity with size. For crystals formed by aggregation rather than accretion, the terminal velocity does not tend to increase with the size of the crystal, because of the crystal's large surface-to-mass ratio. The drag force on a crystal increases in proportion to its mass, and hence an unrimed crystal 2 mm in diameter may have a terminal velocity of $0·5 \ m \ s^{-1}$, whilst a rimed ice pellet of similar diameter would have a terminal velocity of $1·8 \ m \ s^{-1}$. Therefore, an ice pellet falling through a cloud of high water content would grow very quickly be accretion. Riming, or accretion, can also cause a rapid build up of ice on solid surfaces, such as aircraft wings and televison trans- mitters. In such cases, impact freezing will generally occur at all temperatures below $0°C$ and it is especially important between 0 and

$-4°C$, where large numbers of supercooled water droplets exist. In March 1969 a television transmitter in Yorkshire, England, collapsed as a result of excessive loading of rimed ice on its structure.

Figure 4.10 shows a simplified diagram of the precipitation processes occurring in layered clouds associated with a typical middle latitude (extra-tropical) cyclone. The cirrus and cirrostratus clouds are composed entirely of ice crystals, whose growth is dependent on the number of ice nuclei and the amount of water vapour available. The freezing process in cirrus clouds produces ice splinters which thereby increase the number of ice nuclei. These ice splinters have fall speeds of approximately $0 \cdot 5$ m s^{-1} and they can, in turn, seed the supercooled clouds below. Ice crystals do not evaporate as readily as water droplets and they may fall through layers of clear air up to 500 m thick before finally evaporating. This can occasionally be observed in cirrus, where fall streaks give a characteristic hook shape to the cloud. In the middle-level clouds, such as altostratus, the clouds are mixed, and both Bergeron and accretion processes are important. At distances of $1 \cdot 5$ km or less above the freezing level aggregation becomes the dominant process and, depending on the temperature of the lower layer of the atmosphere, precipitation will fall as either rain or snow.

The growth of hailstones is an extreme example of the accretion process and it occurs in deep convective clouds, which have both a high water content and very strong updraft. Under exceptional circumstances, hailstones as large as 13 cm in diameter (about the same size as a grapefruit) have been recorded, although most hailstones are less than 1 cm across. Even in the U.K., hailstones as large as 7 cm have been recorded in summer storms. Hailstones are formed in only a small region of a cumulonimbus cloud on the forward side of the updraft region, as

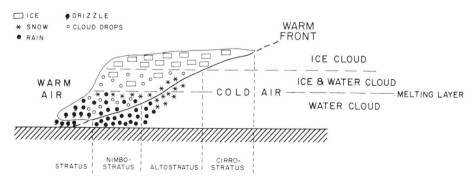

Figure 4.10. The Norwegian model of the microphysical structure of a warm front. (Reproduced from Atmospheric Science, *by J. M. Wallace and P. V. Hobbs, p. 255, figure 5.32, with permission of Academic Press.)*

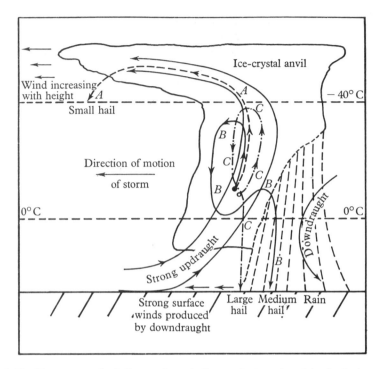

Figure 4.11. The structure of a hailstorm. Large hail occurs in the region of the cloud where recircula-tion in the updraft is possible. (Reproduced from Clouds, Rain and Rainmaking, *By B. J. Mason, p. 94, figure 24, with permission of Cambridge University Press.)*

shown in figure 4.11. The hailstones usually occur for a few minutes in the course of a storm which may last for up to 1 hour. When a thin section of a hailstone is viewed in transmitted light, a banded appearance is usually seen, with alternating layers of clear and opaque ice. This banded structure is the result of the impact of supercooled water droplets at different heights within the cloud. In the upper levels of the cloud, where the temperature range is from -20 to $-40°C$, water droplets are not numerous, freezing occurs on impact and air is trapped between the layers, thus giving a milky appearance. At lower levels in the cloud, where the temperature range is from -20 to $0°C$, there are more water droplets, and the hailstone collects water droplets at too fast a rate to be able to dissipate the latent heat of fusion by conduction to the air as required if spontaneous freezing is to occur. The surface of the hailstone becomes wet and it eventually freezes to form a layer of clear ice. A hailstone will often require a number of cycles within the cloud to grow to a reasonable size and for this to occur the updraft must exceed $10 \, \mathrm{m\,s^{-1}}$. As a general rule, the maximum updraft in the upper troposphere is equivalent to the final terminal velocity of the hailstone. A hailstone 1 cm in diameter has a terminal fall speed of 25 m s^{-1}, whilst a 10 cm hailstone has a fall speed of 65 m s^{-1}. The latter size of hailstone would have a mass of about $0 \cdot 5$ kg and would be able to shatter glass, break roof tiles and indent cars and caravans.

4.6. Macroscopic processes in cloud formation

One advantage that the meteorologist has over the oceanographer is his ability to see the formation of clouds and to use his direct observations to make deductions about atmospheric motions and processes. From a microphysical point of view, it is possible to distinguish three types of cloud. At temperatures below $-39°C$ all clouds are composed of ice crystals. These clouds are known as 'cirrus' and they tend to have fuzzy edges, because of the slow evaporation of ice crystals ejected from the cloud, as discussed in Section 4.5. These cirrus clouds are optically thin and they allow solar radiation to penetrate to lower levels in the atmosphere. The alignment of these clouds gives information on the direction of the wind and the fall streaks allow an estimate of the vertical wind variation to be made. The hexagonal structure of the ice crystals gives rise to preferred directions of scattering and thus to a variety of optical phenomena.

At temperatures between -39 and $0°C$, supercooled water droplets and ice crystals co-exist and in this temperature range the clouds are, therefore, termed 'mixed clouds'. At temperatures colder than $-30°C$, the water droplets are outnumbered by the ice crystals and so mixed clouds tend to behave optically like pure ice clouds at these low temperatures. Between -10 and $0°C$, ice crystals are few in number and therefore these clouds behave optically as if they were water clouds. Typical mixed clouds are altocumulus and altostratus, which form at mid-levels in the troposphere. In thin mixed clouds, with small water droplets, coloured diffraction rings can appear round the Sun or Moon, forming a corona.

At temperatures above $0°C$, clouds are composed solely of water droplets. Such clouds include small cumulus, stratus and stratocumulus and they exist at low levels in the troposphere. Water-droplet clouds tend to have distinct edges, due to the rapid evaporation of ejected water droplets in an unsaturated environment. Such sharp delineation is often seen in growing cumulus.

However, the microscopic classification of clouds has its limitations in ascertaining cloud processes since, for example, it does not distinguish between cumulus- and stratus-type clouds. These two types of cloud, which are the basis of the descriptive classification of clouds formulated in the nineteenth century by Luke Howard, give information about the stability of the atmosphere. Layered, or stratus, clouds form in a locally stable environment, whilst lumpy, or cumulus, clouds are indicative of vertical instabilities. Low-level stratus-type clouds can be formed by:

(i) Advection of warm, moist air over a cooler surface. This may cause advection fog, particularly over cold currents and snow-covered surfaces.

(ii) Radiation cooling of a moist air mass during its transition from low to high latitudes.

(iii) Gentle uplift of a stable air mass, either by a topographic feature or in a large frontal region associated with an extratropical cyclone.

On the other hand, cumulus-type clouds are formed in a locally unstable environment produced by advection of cool, drier air over a warm (moist) surface. In summer, a stable air mass may be rendered unstable in the lower 1 km by surface heating over land. This gives rise to small cumulus, known as fair-weather cumulus, providing that the convection reaches the condensation level. Over the ocean, polar air masses moving equatorwards over warmer surface waters can give rise to convection. The depth of the convection depends upon the dynamical motions maintaining the stability of the troposphere. For example, in an atmosphere of weak vertical stability, such as occurs in the rear of an extra-tropical cyclone or in the ITCZ, convection may extend from the sea surface to the top of the troposphere. In such cases, deep convective cumulonimbus clouds will form. These clouds will consist of water droplets in the lower troposphere and predominantly of ice crystals in the cirrus anvil in the top of the troposphere. In the trade-wind region, or in a middle latitude anticylone, where the air has strong vertical stability, the convection will be limited and cumulus will form in the moist boundary layer. However, progressive warming of a trade-wind air mass as it crosses the ocean may be sufficient to produce convection 2–3 km deep, which is sufficient to produce showers. A less commonly observed convective cloud is termed 'arctic smoke' and it is formed by the movement of a very cold air mass over a warmer sea surface. The layer of air adjacent to the sea is strongly unstable and transfers heat and, more importantly, water vapour into the air mass and this water vapour condenses on mixing with its cold environment. This process quickly warms the arctic air mass.

Instability may be induced not only by surface heating but also by the advection of colder air over a warm air mass. This type of instability often occurs in summer, when a cooler maritime air mass is advected over a very warm surface air mass of continental origin. A notable feature of this type of instability is the appearance of lines of convective cloud in the middle troposphere, which are known as castellanus clouds because of their 'castle-like' appearance.

Unstable and stable air masses may co-exist and produce a variety of mixed clouds. As noted earlier, a stable air mass may be rendered unstable by surface heating which, in turn, leads to the development of cumulus. If there is sufficient moisture present and if the surface heating is strong enough, the cumulus may develop to a depth of 1 km, which may be enough for the development of an isolated light shower. However, the stability of the air above the cloud may be strong enough to limit further upward development of the cloud which will, therefore, spread laterally. This cloud, known as stratocumulus, will often produce a complete layer of cloud cover, which will persist until the surface heating decreases or until sufficient dry air from above the inversion is mixed into the clouds to promote evaporation. This type of cloud often has a diurnal cycle, with maximum thickness occurring in the early afternoon, and usually clearing at dusk, when the surface heating is reduced. Alternatively, a layered cloud may break up into convective-type cloud because of solar radiational heating. As discussed in Chapter 2, solar radiation penetrates a cloud to a considerably deeper depth than long-wave planetary radiation. Thus the top of a cloud will cool by long-wave radiative emission into space whilst the bulk of the cloud mass will heat up by solar radiation absorption. This process results in an increase in the vertical temperature gradient within the cloud which, in turn, leads to instability. This phenomenon is more often observed in thin clouds, like altostratus or cirrostratus, which are transformed into altocumulus and cirrocumulus.

The mixing of air masses is another important macroscopic cloud process. The importance of the mixing of cloudy air with dry air to dissipate cloud has already been mentioned, but if two air masses close to saturation mix, clouds may form. The reason for this may be understood by referring to figure 4.12. If air mass A and air mass B are mixed, then the resultant air mass will lie along the line *AB* and its exact location will depend on the relative properties of the two air masses. The saturation mixing ratio does not vary linearly with temperature and so the points lying along the line *A'B'* will represent saturated or supersaturated air and thus condensation would be expected. Only for mixtures lying on the lines *AA'* and *BB'* will the air remain unsaturated. A common example of this phenomenon is the emission of hot, moist exhaust gases from a car or an aeroplane into a cold, moist environment. In this case, because of the low environmental temperature, only a small amount of water vapour is required for condensation.

Vertical mixing of air masses may also result in cloud formation. One particular example is the formation of stratocumulus at the base of the inversion in an anticyclonic system. In this case, mixing of cool moist surface air with warm dry air aloft will result in a decrease in temperature, but an increase of mixing ratio at the top of the layer. The

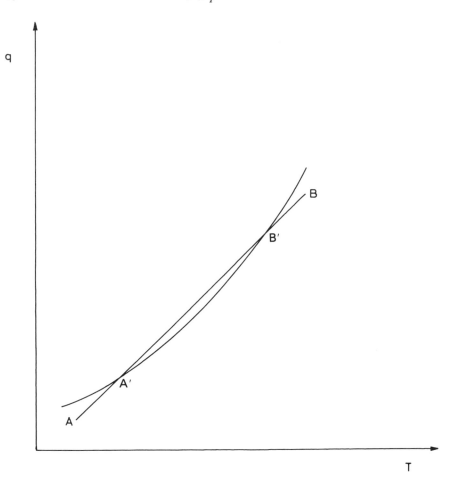

Figure 4.12. Mixing of two unsaturated air masses (A, B). A mixture lying on the line A'B' will become saturated.

converse will occur in the lower layer, thus resulting in condensation in the upper layer. This process is often responsible for the development of stratocumulus over the sea, when the sea temperature is lower than the temperature of the air above it. In a winter anticyclone, over a moist land surface, a thick layer of stratocumulus may persist for many days, giving rise to conditions known as 'anticyclonic gloom'. Clearly, the formation of this low-level cloud is dependent on a number of processes. The upward flux of water vapour from the surface and vertical mixing will tend to enhance cloud formation, whilst the downward mixing of warmer, dry air from above the inversion will tend to dry out the boundary layer and dissipate the cloud.

Vertical mixing is also important in the development of surface fogs. These may develop by long-wave radiational cooling of the land at night which, if sufficient, will lower the ground temperature to the dew point of the overlying air. Providing the air is still, only the surface layers of air will reach the dew point and a shallow layer of fog will form. However, with vertical mixing, the depth of air cooled to the dew point will increase and hence a thicker layer of fog will develop. Another example is the formation of advection fog, formed by the flow of warm, moist air over a cold surface, such as a cold ocean current or a snow-covered land surface. In this case, vertical mixing will tend to enhance the depth of the fog.

The third, and most important, contribution to the formation of clouds is adiabatic cooling. A spectacular example is the formation of funnel cloud in a tornado vortex. The low pressure of such a vortex is often sufficient to adiabatically cool the air to the condensation point. A fall of 50 mb in the pressure will cause an adiabatic cooling of $4°C$ at the surface, but by far the most common cause of condensation is adiabatic cooling by vertical uplift. This vertical motion may be caused by the forced lifting of an air mass over a topographical feature, by convection (as discussed previously) or by large-scale ascent associated with frontal or synoptic scale features. The lifting of air masses over topography may give rise to relatively strong vertical motions of $1 \, \mathrm{m \, s}^{-1}$ and to the development of 'local' clouds over the topographic feature. In a stable environment, wave clouds may form over the mountain and downstream of the mountain for up to 100 km. Clouds formed by large ascents in fronts tend to be associated with weaker vertical motions of approximately $0 \cdot 1 \, \mathrm{m \, s}^{-1}$ and therefore the clouds tend to be layered. However, regions of stronger uplift, associated with rain bands in a frontal system, may produce more convective-type cloud.

5

Global budgets of heat, water and salt

5.1. The measurement of heat budgets at the surface

The net heat flux, Q_T, across a horizontal surface is given by

$$Q_T = Q_S(1 - \alpha) - Q_N - Q_E - Q_H \qquad (5.1)$$

where Q_S is the downward solar radiation flux, α is the surface albedo, Q_N is the net upward or long-wave planetary radiation flux, Q_E is the heat lost by the surface due to the latent heat of evaporation and Q_H is the upward turbulent heat flux. Thus four separate quantities must be determined in order to calculate the total heat flux across an ocean or land surface, and the method used over land and sea to make the measurements over limited areas will now be discussed. The net radiation flux, Q_R, is given by $Q_R = Q_S(1 - \alpha) - Q_N$.

The solar radiation flux, Q_S, can be determined using a solarimeter, which measures both the direct and diffuse radiation between $0 \cdot 35$ and $2 \cdot 8$ μm, i.e. over most of the solar spectrum. The solarimeter consists of two elements, one blackened to both solar and long-wave radiation and the other covered with a white paint which has a high short-wave reflectivity. The temperature difference between the two elements, measured by a series of thermocouples, is proportional to the incident short-wave radiation. A glass dome covering the sensor elements gives the instrument protection from the weather, and it can be mounted on a ship and give reliable readings for long periods of time. The accuracy of the instrument is about 2%.

The net long-wave flux, Q_N, can be determined from measurements of the total net radiation and the solar radiation. A net radiometer consists of two blackened thermopiles, one pointing upwards and the other pointing downwards. The voltage difference between the two thermopiles is a measure of the net radiation across the surface, which may thus be determined with an accuracy of about $2 \cdot 5$%. The thermopiles are enclosed in a polythene dome which transmits both long- and short-wave radiation.

The dome is ventilated with dry nitrogen gas to avoid water vapour absorption and condensation. Measurements of net long-wave radiation are obtained routinely at many meteorological stations but measurements at sea are more difficult. It is necessary to avoid contamination by reflections from the ship and thus a boom is required to obtain uncontaminated fluxes. However, providing that the sea surface temperature is known, the upward flux of long-wave radiation can be calculated from Stefan's law, $E = \epsilon \sigma T^4$, with an emissivity, ϵ, of $0 \cdot 97$, and thus only the downward long-wave flux from the sky needs to be measured. This can be done using an upward-looking radiometer.

The most accurate determination of the latent heat flux, Q_E, and the upward turbulent heat flux, Q_H, is by the eddy correlation method. Over both land and sea, the upward flux of heat (and water vapour) is caused by turbulent motions close to the surface. Figure 5.1 shows the fluctuations of vertical velocity and temperature for a 2 min record at 23 m over land. Despite the randomness of the fluctuations, it can be seen that high temperatures are generally associated with the rising motion, whilst low temperatures are associated with the descending motion. In this situation, a mean vertical gradient of temperature exists with high temperature close to the surface. Fluctuations in vertical velocity, associated with convection or surface friction, cause an exchange of parcels of air across the gradient and this, in turn, gives rise to an upward flux of heat. The cor-

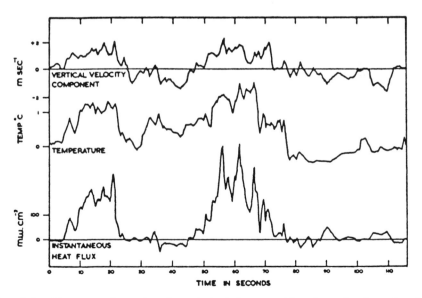

Figure 5.1.   Temperature fluctuations, θ, vertical velocity, w, and heat flux, $\rho C_P w \theta$, at 2 m above the Earth's surface. (Reproduced from Turbulent Transfer in the Lower Atmosphere *by C. H. B. Priestley, p. 67, figure 18, with permission of University of Chicago Press.)*

relation coefficient between the vertical velocity and temperature gives an accurate measurement of the heat flux at the level of instrumentation. Providing that there is no loss of heat between the surface and the instrument, the correlation coefficient will give a measurement of the surface turbulent heat flux. In a similar manner, the correlation coefficient between the vertical velocity and humidity fluctuations allows a determination of the latent heat flux. The turbulent heat flux is given by:

$$Q_H = \varrho C_p \overline{w\theta} \tag{5.2}$$

and the latent heat flux by:

$$Q_E = L\varrho \overline{wq} \tag{5.3}$$

where the overbar denotes a time average of the instantaneous values of the vertical velocity, w, the mixing ratio, q, and the potential temperature, θ, over a period of 10–60 min.

This method requires sophisticated and accurate instrumentation in order to measure the high frequency variations in w, q and θ. In one 10 min average, about 10,000 observations are required to obtain reliable fluxes. Over the sea, there are more difficulties because of the ship's movements and the contamination of the instruments by salt spray. For these reasons, the above technique is only practicable for the determination of latent heat and turbulent heat fluxes from a research ship, for limited periods and in light to moderate wind conditions. However, the method is very useful in the calibration of other, simpler formulae which are more widely used.

More common methods of measuring the latent heat and turbulent heat fluxes rely on the measurement of the mean vertical gradients of humidity and temperature which can be obtained, for example, using a standard ventilated psychrometer. By analogy with heat conduction in solids, the flux of latent heat can be written as:

$$Q_E = -\varrho L K_E \frac{\Delta q}{\Delta z} \tag{5.4}$$

and the flux of turbulent heat as:

$$Q_H = -\varrho C_p K_H \frac{\Delta \theta}{\Delta z} \tag{5.5}$$

where $\Delta q/\Delta z$ and $\Delta\theta/\Delta z$ are the vertical gradients of the water-vapour mixing ratio and the potential temperature, respectively. The coefficients K_E and K_H are known as the eddy exchange coefficients and have to be determined for each set of observations, unlike the thermal conduction coefficients which remain constant over reasonably wide ranges of temperature and pressure. In order to calculate the exchange coefficients, an equation for the tangential stress, τ, must be introduced:

$$\tau = \varrho K_M \frac{\Delta u}{\Delta z} \tag{5.6}$$

where K_M is the eddy exchange coefficient for momentum and $\Delta u/\Delta z$ is the vertical gradient of the wind speed. Although the eddy coefficients all vary with height above the surface and with the stability of the atmosphere, it is reasonable to assume that, at a given height and time, the mixing processes for humidity, temperature and momentum will be similar and vertical eddy circulations will carry parcels of momentum, temperature and humidity simultaneously. In this case, $K_E = K_H = K_M = K$ and substituting for K from equation 5.6 into 5.4 and 5.5:

$$Q_E = L\tau \frac{\Delta q}{\Delta u} \tag{5.7}$$

and

$$Q_H = C_p \tau \frac{\Delta \theta}{\Delta u} \tag{5.8}$$

From work on aerodynamic boundary layers, the surface stress, τ_0, is known to be given by:

$$\tau_0 = \varrho C_D U_{10}{}^2 \tag{5.9}$$

where C_D is the drag coefficient and U_{10} is the wind speed 10 m above the surface.

For a surface stress, the wind speed at the surface is zero and so $\Delta u = U_{10}$. In addition:

$$\Delta q = q_A - q(\theta_S) \tag{5.10}$$

and

$$\Delta \theta = \theta_A - \theta_S \tag{5.11}$$

where θ_A and q_A are the potential air temperature and mixing ratio, respectively, 10 m above the surface, and θ_S is the sea surface temperature. $q(\theta_S)$ is the surface mixing ratio.

Substitution of equation 5.9 into 5.7 and 5.8, and using 5.10 and 5.11

yields:

$$Q_E = -\varrho L C_D U_{10}(q_A - q(\theta_S)) \tag{5.12}$$

$$Q_H = -\varrho C_p C_D U_{10}(\theta_A - \theta_S) \tag{5.13}$$

Over the sea surface, $q(\theta_S) = q_s(\theta_S)$, where q_s is the saturation mixing ratio. These relationships are most applicable to the computation of fluxes over the sea surface. They require only observations of U_{10}, q_A, θ_A and θ_S and these can be obtained from ships' reports as a matter of routine. However, they have the disadvantage that the drag coefficient, C_D, is not a constant, being dependent on both vertical stability and on wind speed. It is fortunate that, over large areas of the ocean, the atmospheric boundary layer near to the sea surface is often close to neutral stability and that the dependence on wind speed only becomes noticeable at high wind speeds. Comparisons between the eddy method and the aerodynamic equations have shown that the latter are accurate to about 10% in moderate wind conditions over the ocean.

Over the land, the aerodynamic equations 5.12 and 5.13 are not as useful because the drag coefficient will vary with the type of surface. In this case, in order to use equations 5.7 and 5.8, τ has to be determined from the wind profile from the surface to about 10 m, which, in neutral stability conditions is given by:

$$U = \frac{U_*}{k} \ln\left(\frac{z}{z_0}\right) \tag{5.14}$$

where U_* is the friction velocity and k is the von Karman constant which has a value of $0 \cdot 4$. z_0 is the roughness coefficient. The friction velocity is defined by:

$$\tau = \varrho U_*^2 \tag{5.15}$$

Thus, from a vertical profile of U, the gradient will determine the friction velocity, U_*, and thence the stress, τ. The friction coefficient, z_0, can also be determined from the velocity profile. Typical values of z_0 vary between $0 \cdot 01$ cm over smooth sand to about 100 cm over scrub country.

Apart from the use of the aerodynamic method, all of the previous methods rely on instrumentation which is only available on research ships, weather ships and some automatic ocean buoys. If it is assumed that there are, at most, 100 specially equipped research vessels at sea at any one time (and this is probably a considerable overestimate), then, provided that they were evenly distributed, one observation would be made for each 2000 km square of ocean. This is an unsatisfactory data set for the consideration of surface fluxes on a global scale. The global maps of surface fluxes in the scientific literature are compiled from the merchant vessel observations rather than those from research ships. As noted, the bulk aerodynamical method is suitable for estimating the turbulent heat flux and the latent heat flux from standard weather reports, but, even here, there are problems. The height of the wind observations will vary from ship to ship, wet-bulb thermometers can be contaminated by salt spray and may, therefore, give a systematic error in humidity if they are not regularly maintained. In addition, the water temperature is measured at the engine intake point, which may be a few metres below the sea surface, and the recorded temperature may be systematically different from the sea surface temperature, particularly in low wind conditions when vertical mixing is reduced.

The problems posed by the global measurement of solar radiation and long-wave radiation are rather more difficult to overcome than those mentioned above, because of the variable effect of cloud. In clear sky conditions, provided that the aerosol is low, both the solar radiation and the long-wave radiation can be determined from empirical formulae. The solar radiation impinging on the top of the atmosphere can be calculated for a given latitude and time of year, as demonstrated in Chapter 3. Allowance can be made for the depletion of the solar radiation by atmospheric gases and a 'standard' maritime aerosol to obtain a value for the solar radiation incident at the surface. However, recent research has shown that horizontal variations of aerosol can cause variations of as much as 5% in the solar radiation. The simplest formulae which include the effect of cloud are of the form:

$$Q_S = Q_C(1 - \alpha_c C) \tag{5.16}$$

where Q_C is the daily averaged clear-sky solar radiation, C is the cloud fraction and α_c is typically $0 \cdot 7$. The use of this formula, whilst reasonable for the determination of monthly averages, can give errors of 50% in estimating daily averages. Substantial improvements on equation 5.16 can be obtained by using information on the type of cloud and the height of the cloud base, which are reported regularly in standard meteorological observations by ships.

Lumb (1964) showed that a more accurate formula for estimating hourly values of solar radiation is given by

$$Q = Q^*(a_1 s + b_1 s^2) \tag{5.17}$$

where s is the sine of the solar altitude and Q^* is the solar constant. a_1 and b_1 are coefficients which depend on nine categories of cloud conditions, which involve type, amount and cloud-base height. This formula is accurate to $\pm 10\%$ for estimating the daily averaged solar radiation, between 45 and 65°N. Recent work in the north-eastern Atlantic Ocean has shown that 12 day average solar radiation estimates from equation 5.17 agree with direct measurements to within 5%.

Net long-wave radiation under clear skies can be estimated from a formula proposed by Brunt (1932):

$$Q_{NC} = \epsilon \sigma T_S^4 (0 \cdot 35 - 0 \cdot 05 \sqrt{e_a}) \qquad (5.18)$$

where ϵ is the emissivity of the sea surface ($\epsilon = 0 \cdot 97$), T_S is the surface temperature, measured in degrees Kelvin, σ is the Stefan–Boltzmann constant and e_a is the vapour pressure of the atmosphere at 10 m. The dependence on humidity is due to the re-emission of the long-wave radiation back to the surface by atmospheric water vapour. The majority of the re-emission comes from the first 100 m above the surface and the humidity at 10 m can, therefore, be used to estimate the downward long-wave radiation. There are two points to note about formula 5.18. First, over the observed range of sea surface temperatures, the net long-wave flux, assuming saturated air and clear skies, only varies from $70 \mathrm{~W} \mathrm{~m}^{-2}$ at 271 K to $23 \mathrm{~W} \mathrm{~m}^{-2}$ at 300 K, which is much smaller than the latitudinal variation of solar radiation. Second, the net long-wave radiation decreases with increasing temperature, because the downward long-wave radiation, which is proportional to the vapour pressure, increases more quickly with surface temperature than the upward radiation emitted by the surface. As with solar radiation, a cloud factor may be introduced in the form:

$$Q_N = Q_{NC}(1 - \alpha_c C) \qquad (5.19)$$

where C is the cloud fraction and $\alpha_c = 0 \cdot 8$. Some formulae allow a variable α_c, dependent on the cloud type. For example, for low-level cloud like stratocumulus, $\alpha_c = 0 \cdot 9$, whilst for high-level cloud such as cirrus, $\alpha_c = 0 \cdot 25$. It is noted that, with total cover of low-level cloud, the net long-wave flux is approximately 10% of the net flux in clear sky conditions, and so the cloud cover will further reduce the latitudinal variation of the long-wave radiation budget. Net radiation can be estimated from equation 5.18 with an accuracy of about 10%.

5.2. *Observations of surface heat budgets*

Table 5.1 shows the annually averaged components of the heat budget for land and ocean. It is noted that over the annual cycle, the net heat flux into the land will be close to zero because the heat gained in the summer

Table 5.1. Annual surface heat budgets (watts per square metre) over ocean and land. Q_R is the net radiation budget, Q_E is the latent heat of evaporation, Q_H is the upward conduction of heat and Q_T is the net heat flux into the ocean.

Latitude zone	Ocean				Land		
	Q_R	Q_E	Q_H	Q_T	Q_R	Q_E	Q_H
50–70°N	35	48	21	− 34	33	22	11
30–50°N	90	93	18	− 21	70	31	39
10–30°N	155	137	10	8	94	33	61
10°S–10°N	154	110	5	39	96	66	30
30–10°S	143	137	8	− 2	96	46	50
50–30°S	93	90	11	− 8	69	33	36

period will be lost to the atmosphere during the winter. Over the ocean, there is a net gain of heat between 20°S and 20°N and at higher latitudes there is a net heat loss. These heat deficits and surpluses must be compensated for by a transfer of heat from the equatorial zones to the middle and high latitudes via ocean currents.

The annual radiation budgets at lower latitudes generally show more heat going into the ocean than into the land. This observation is related to two factors. First, the albedo of the sea surface is much lower, on average, than that of the land at corresponding latitudes, especially the sub-tropical deserts, and therefore more solar radiation is absorbed by the ocean. Second, the surface temperatures of the tropical ocean are lower than those observed over land and therefore the net long-wave radiation losses are less. Over the ocean, the largest net loss of heat is by evaporation; the turbulent heat conduction accounting for only 10% on average of the net heat loss. In the equatorial zone, Q_H contributes only 5% to the total heat loss whilst at higher latitudes it contributes about 30%. Over land, both latent heat and heat conduction contribute equally to the overall net heat loss. In the sub-tropical deserts heat conduction dominates over evaporation whilst the opposite is true in the equatorial zone and over the wetter middle latitudes.

To gain an impression of the longitudinal variations in surface heat budgets over the ocean, the annual average components are shown in figure 5.2 for the North Atlantic, calculated from 25 years' observations by merchant vessels. Large evaporative heat losses of over 200 W m^{-2} are seen over the Gulf Stream and these losses are twice as large as the evaporative losses in the sub-tropics. Furthermore, large turbulent heat losses are seen in the western Atlantic. In contrast, the radiation budget (Q_R) is rather zonal in character. Net heat gain occurs in the southern and eastern Atlantic, whilst net heat loss occurs in the northern and western

(a) Q_T, Wm^{-2}

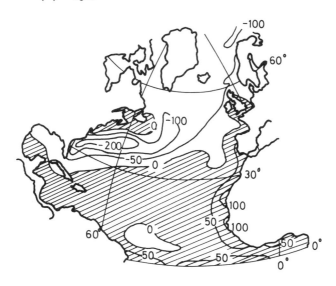

(b) Q_H, Wm^{-2}

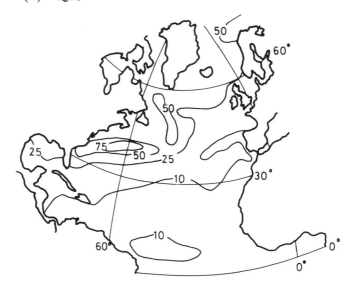

Figure 5.2. (a) Annual average heat gain, Q_T ($W\,m^{-2}$), for the North Atlantic. (b) Annual average turbulent heat flux, Q_H ($W\,m^{-2}$). (c) Annual average evaporation, E ($cm\,year^{-1}$). Evaporation rate should be multiplied by 1·26 to get approximate heat flux, Q_E ($W\,m^{-2}$). (d) Annual net radiation exchange, Q_R ($W\,m^{-2}$). (Reproduced from A. F. Bunker, Monthly Weather Review, **104**, *1130 and 1131, figures 4a, 4b, 4c and 4d, with permission of the American Meteorological Society.)*

(*c*) Q_E, cm year^{-1}

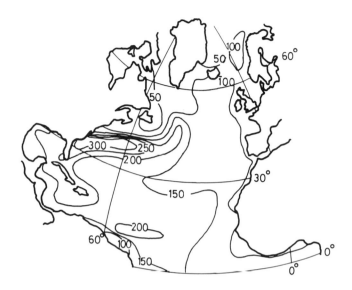

(*d*) Q_R, Wm^{-2}

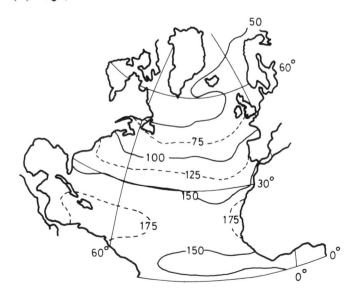

Atlantic. The large heat losses in the western Atlantic are caused by a combination of the high sea surface temperatures and the frequent out-breaks of cold, dry air from the North American continent in winter. In the eastern Atlantic, the sea surface temperature is lower, specially in the cold upwelling zones off north-west Africa, and therefore evaporation and turbulent heat flux are reduced. The cold upwelling zone along the equator can easily be identified by the suppression of evaporation and the large net heat flux into the ocean.

It will now be shown how the distribution of the net heat flux across the ocean surface can be related to the transport of heat in the ocean. First, consider the heat budget for a slab of ocean, extending across an ocean basin and over the whole depth. For the conservation of heat:

$$S_O + T_{i+1} - T_i = Q_T \tag{5.20}$$

where S_O is the change of storage of heat in the slab which is proportional to the rate of change of mean temperature. T_i and T_{i+1} are the net horizontal transports of heat into and out of the slab. Their difference is known as the heat flux divergence. Positive values indicate heat divergence and negative values indicate heat convergence. Equation 5.20 indicates that a net heat flux, Q_T, into the ocean must be accompanied by either a change in the heat storage or by the net advection of heat out of the slab. To obtain the heat transport across the boundary of each slab, it is assumed that the northward flux across the strip closest to the North Pole is zero. Hence by summing consecutive strips, T_{i+1} can be obtained:

$$T_{i+1} = Q_T - S_O + T_i \tag{5.21}$$

The small contributions of frictional dissipation by tides and currents, by precipitation and by radioactive decay in the Earth's interior have all been neglected. All of these factors are likely to contribute less than 1 W m^{-2} to the net heat flux. In the ocean, and in marked contrast to over the land, the storage component, S_O, can be as large as the net heat flux, Q_T, during the seasonal cycle because of the important role played by the ocean in increasing the thermal inertia of the atmosphere–ocean system. Figure 5.3 shows a map of S_O, $T_{i+1} - T_i$ and Q_T for the Northern Hemisphere during the seasonal cycle. Comparison of S_O and Q_T shows the large accumulation of heat in the middle latitudes during the summer and the release of that stored heat during the winter. However, it is noted

Figure 5.3. (a) Vertical heat flux into the ocean, Q_T, (b) storage, S_O, (c) heat divergence and (d) heat transport, T_i, in the Northern Hemisphere. (Reproduced from A. H. Oort and T. H. vonder Haar (1976), J.P.O., 6, 789–795, figures 5, 8, 9 and 13, with permission of the American Meteorological Society.)

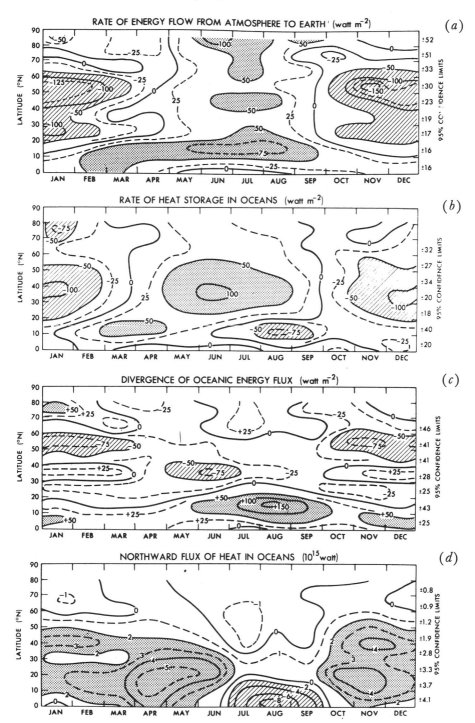

(a) RATE OF ENERGY FLOW FROM ATMOSPHERE TO EARTH (watt m⁻²)

(b) RATE OF HEAT STORAGE IN OCEANS (watt m⁻²)

(c) DIVERGENCE OF OCEANIC ENERGY FLUX (watt m⁻²)

(d) NORTHWARD FLUX OF HEAT IN OCEANS (10¹⁵ watt)

that the middle latitude heat loss in winter is greater than the heat accumulated in summer, and so there must be a poleward flux of heat to maintain the balance. This flux is, on average, about 2–3 petawatts (1 petawatt = 10^{15} watts) at 25°N.

The change of heat storage occurs mainly over the upper 100 m of the ocean, except in high latitudes of the Atlantic Ocean where it occurs over 500 m. It is also noted that, averaged over the seasonal cycle, the storage term tends to cancel out, and it is therefore reasonable to assume that, over a long period of observations, say 25 years, the change in storage will become small and a balance between the transport of heat and the net heat flux across the surface will occur (equation 5.21). This is a steady-state assumption which implies that, in the long term, the mean ocean temperature is a constant. It is not possible to test this assumption in the ocean because of insufficient high-quality data over a long enough period of time. However, long-term trends in the sea surface temperature have shown increases of 0·5 K over the last 100 years. If this trend is occurring over the whole depth of the ocean, then an estimated net heat flux into the ocean of $2·5\ \mathrm{W\,m^{-2}}$ is required. (This is likely to be an overestimate because the mixing time scale for the whole ocean is between 200 and 1000 years, and therefore this warming trend would not be shown by the whole mass of the ocean.) Compared with the net heat divergence shown in figure 5.3, the steady-state assumption yields an error which is generally less than 10%.

Figure 5.4 shows the long-term mean meridional transport of heat in the ocean, estimated by applying equation 5.20 to each ocean basin. This map gives only a rough picture of the heat transport in the ocean because of the errors in estimating the net heat flux into the ocean, as discussed in Section 5.1, and because of the uncertainty about long-term changes in storage. Of particular interest, however, is the difference between the Atlantic and Pacific Oceans. In the Atlantic Ocean, there is a northward heat flux throughout the whole ocean which is maintained by a net heat flux from the Indian and South Pacific Oceans. In contrast, in the Pacific Ocean there is a poleward heat flux in both hemispheres with little or no net heat flux across the equator. It is also noted that the Arctic Ocean has a substantial heat flux from the North Atlantic Ocean but gains no significant heat from the Pacific Ocean through the Bering Straits.

An overall conservation equation for the entire atmosphere–ocean can be determined in a similar way to equation 5.20. A box extending from the base of the ocean to the top of the atmosphere is shown in figure 5.5. The net heat flux, R, into the top of the atmosphere is simply the sum of the net solar radiation and net outgoing planetary radiation, which can be determined from satellite observations. The divergence of atmospheric energy, i.e. internal energy, potential energy and latent heat flux, is represented by T_A, the atmospheric storage term by S_A, the land storage

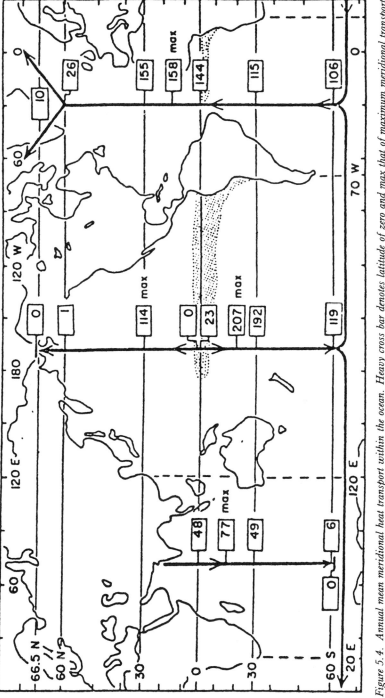

Figure 5.4. Annual mean meridional heat transport within the ocean. Heavy cross bar denotes latitude of zero and max that of maximum meridional transport, with numbers indicating amounts in units of 10^{15} W. Stippling marks area with $Q_T > +50\ Wm^{-2}$ and broken lines show meridians used as boundaries between oceans in high southern latitudes. (Reproduced from H. Stommel (1980), Proc. Nat. Acad. Sci., USA, 77, 2378, figure 1, with permission of the National Academy of Sciences, USA.)

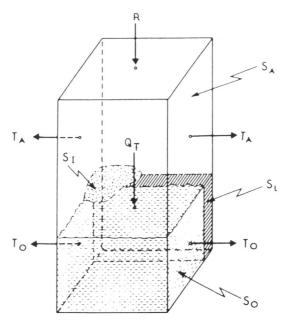

Figure 5.5. Schematic of the different terms in the Earth's energy budget. R *is the net radiation flux at the top of the atmosphere and* Q_T *is the net surface heat flux.* S_A, S_I, S_L *and* S_O *are the rates of storage energy in atmosphere, snow and ice, land and ocean, respectively.* T_A *and* T_O *are the horizontal divergences of energy in the atmosphere and the ocean. (Reproduced from A. H. Oort and T. H. vonder Haar (1976), J.P.O., 6, 783, figure 1, with permission of the American Meteorological Society.)*

and ice storage by S_L and S_I respectively, the ocean storage by S_O and the divergence of the ocean by T_O, then:

$$R = T_A + T_O + S_A + S_O + S_L + S_I \qquad (5.22)$$

If, over a long period of time, the storage terms are assumed to be negligible, then:

$$R = T_A + T_O \qquad (5.23)$$

The term $T_A + T_O$ is the total energy divergence of the atmosphere–ocean system. In a similar process to the determination of the poleward transport of heat in the ocean, equation 5.23 can be used to determine the total poleward transport by atmosphere and ocean. Figure 1.6 shows that poleward of $40°$ there is a net deficit of radiation, because the outgoing long-wave radiation exceeds the incoming solar radiation, and so a poleward transport of energy is required to maintain the balance. The atmospheric energy transport can be obtained by sub-tracting the ocean transport from the total transport. In low latitudes,

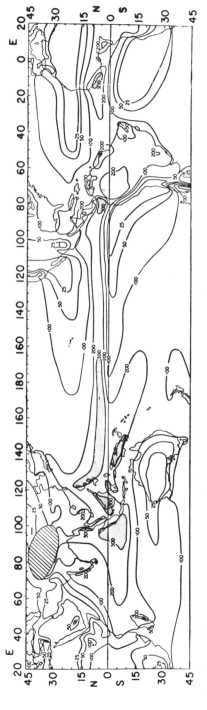

Figure 5.6. Mean annual precipitation (cm). Shaded regions have more than 300 cm year⁻¹. (Reproduced from Climate and Weather in the Tropics, by H. Riehl, p. 87, figure 34, with permission of Academic Press.)

the ocean transport is larger than the atmospheric transport and it reaches a maximum at 25°N. However, in middle latitudes and high latitudes the atmospheric transport is much larger, accounting for over 70% of the total transport at 40°N, and virtually all the transport poleward of 60°N. A similar pattern has been deduced for the Southern Hemisphere.

5.3. The measurement of the water budget

The vertical flux of water, W, across a horizontal surface, is given by:

$$W = P - E \tag{5.24}$$

where P is the precipitation, including dewfall, and E is the evaporation. Both quantities are usually expressed in units of millimetres per year or centimetres per year. The precipitation map shown in figure 5.6 is derived almost entirely from land-based measurements using rain gauges. At a standard climatological station, when care is used in siting the rain gauge, the precipitation can be determined to an accuracy of approximately 1–2%. All gauges tend to underestimate precipitation because turbulence around the gauge tends to reduce collection efficiency. However, over the oceans, precipitation measurements using rain gauges are neither as frequent nor as accurate as over the land. The reasons for this are related to the problem of wind flow around the ship carrying the gauge and, to a lesser extent, the problem of spray. Comparisons between rain gauges mounted on different parts of a ship have shown that differences of 50–100% can be expected, and a number of indirect techniques have, therefore, been employed to measure the rainfall over the ocean. One technique uses the visual observations of rain intensity, which are recorded regularly by merchant and weather ships as part of their meteorological reports, and converts them into numerical values of rainfall intensity. As an example, light continuous rain would be assigned a rainfall intensity of $0\cdot6$ mm hour^{-1}, whilst moderate continuous rain would be given a value of $1\cdot9$ mm hour^{-1}. By this method, annual mean distributions of rainfall can be obtained to an accuracy of about 10%. This technique is probably best suited to higher latitudes where the rainfall is less intense, but more frequent, than in the tropics. Another indirect method is to use radar to measure the back-scattered radiation from large water droplets. The radar can be calibrated by comparing simultaneous measurements of back-scattered radiation with rainfall rates determined by a conventional, land-based, rain-gauge network. This radar technique is not very accurate because the back-scattered radiation is dependent on the size distribution of the droplets. It will not be reliable when there is a large component of frozen precipitation. It is most useful for the detection of intense mesoscale rainfall systems and is limited to a range of about 100 km over land. Over the sea there is a reduced horizon, because of the limited height at which the radar is fixed onto the ship.

In the last decade there has been considerable interest in the use of satellites to measure precipitation. One technique measures the temperature of the cloud tops using infra-red sensors. This can be used to obtain the depth of a cloud which can, in turn, be empirically related to the intensity of precipitation from the cloud. This method is most successful for deep convective clouds and is, therefore, best suited to determine the rainfall distribution over the tropical ocean. However, as with all satellite techniques, its principal value lies in the determination of the spatial distribution of precipitation rather than for obtaining accurate point values. A second satellite method uses a microwave radiometer to measure the small amounts of microwave radiation emitted by the suspended liquid water in the atmosphere. Changes in the liquid water content can be related to the precipitation rate and thus an instantaneous picture of the precipitation pattern may be obtained. Both satellite techniques are limited by the frequency of satellite passes over a particular region; perhaps only one to three passes per day. Rainfall is highly variable in both time and space and so these techniques require further development if they are to be of use in obtaining quantitative climatological precipitation maps similar to that shown in figure 5.6.

The direct measurement of evaporation from a ship is no less difficult than that of precipitation. The simplest method is to measure the rate of evaporation from a pan, filled with salt water and suspended in gymbals. An accurate determination of the water lost can be made by measuring the change in salinity. In practice, this method is unreliable because the evaporation from a pan at deck level will generally be higher than that at sea level due to the vertical gradient of humidity and wind speed over the ocean. To be representative of the evaporation from the ocean surface, the pan would have to be suspended or floated on the surface and this is only possible in the smoothest of seas. Second, solar heating of the pan itself would result in anomalously high evaporation rates. Comparisons of the evaporation at the sea surface with that at deck level have indicated systematic errors of about 50%. Because of all of the limitations and difficulties discussed above, the indirect techniques, such as the bulk aerodynamic method discussed in Section 5.1, are found to have smaller systematic errors, of about 10%, than the pan method. These indirect techniques can be used to make routine calculations from merchant ships' reports. It is noted that an evaporative heat flux of $100 \mathrm{~W~m}^{-2}$, such as observed at the equator, is equivalent to an evaporation of $126 \mathrm{~cm~year}^{-1}$.

5.4. Observations of the water budget

For the reasons discussed above, the distribution of precipitation is only accurately known over land areas. In the Pacific Ocean, the large number of islands in the western and central regions allow a good estimate of the rainfall over the ocean to be obtained, although it is important to realize that islands, especially if they have hills or mountains, will tend to be preferential sites for tropical convection and precipitation. In the tropical South Atlantic the paucity of island sites results in a large uncertainty in the rainfall distribution. Between the equator and $40°S$, there are only seven islands, not all of which record rainfall. However, despite the uncertainty in quantitative measurements over the ocean, the general global distribution (figure 5.6) is probably correct and, in recent years, the previously described satellite techniques have enabled maps of the major precipitation belts to be obtained.

The most significant feature of the precipitation distribution in tropical latitudes is the relationship between the sea surface temperature and the rainfall belts. High rainfall in both the Pacific and Atlantic Oceans occurs along the ITCZ, which correspond closely to the thermal equator.

Of particular note is the very high rainfall over the equatorial regions of South America and Africa and also the maritime continent of Indonesia, which exceeds 300 cm year^{-1}. By contrast, in parts of the eastern tropical and sub-tropical ocean the rainfall is less than 10 cm year^{-1}. This dramatic decrease is particularly noticeable in the south equatorial Pacific, where the Galapagos Islands have a desert climate whilst only 300 km northwards, in the ITCZ, the rainfall may exceed 300 cm year^{-1}. The oceanic desert regions are associated with lower sea surface temperatures which, in turn, are related to the upwelling of cold surface water and equatorial advection of colder water by the sub-tropical anticyclonic gyres.

With the use of climatological cloud distributions, derived from satellite photographs, convergence zones and regions of higher precipitation have been located in all three ocean basins. The most significant is the South Pacific convergence zone which extends south-eastwards from Indonesia into the central South Pacific. Similar cloud zones have been observed over the South Atlantic extending from the Amazon basin, and from East Africa into the southern Indian Ocean. In higher latitudes the precipitation rates are much lower due to the lower water-holding capacity of the cooler atmosphere. The evaporation rate, which is related to both the vertical gradient of humidity and to the wind speed, is large over the tropical trade-wind zones, where both the wind speed and the vertical gradients of humidity are large. In these regions, evaporation reaches 150–200 cm year^{-1}. In the middle latitudes large evaporative fluxes of up to 250 cm year^{-1} are seen over the western North Pacific and western North Atlantic due to advection of dry continental air over a warm ocean surface.

Over the whole globe evaporation must, of course, be in balance with precipitation. The global average precipitation required to balance evaporation is 103 cm year^{-1}, and a horizontal transfer of water must occur to balance regions of water deficit with regions of water surplus. Figure 5.7 shows the latitudinal variation of the components of the water balance. A net surplus of water (i.e. $W > 0$) exists over the equatorial region, associated with the ITCZ, and polewards of $40°$ latitude, associated with the middle-latitude cyclones. In the tropical and sub-tropical zones there is a net deficit (i.e. $W < 0$) as a result of the low precipitation over the ocean and desert regions and the high evaporation in the trade-wind zones. To maintain this latitudinal variation of W, it is necessary to transfer new water in the ocean from the higher latitudes and the equatorial zone into the sub-tropical regions. In the atmosphere a transfer of water vapour in the opposite direction is required to main-tain the regions where precipitation exceeds evaporation. Most of the water vapour is contained in the lower 5 km of the atmosphere and these water-vapour transfers must, therefore, occur in the lower layers of the atmosphere. The trade winds, which occupy the lowest 2–3 km of the atmosphere, are responsible for the transfer of water vapour from the sub-tropical ocean into the ITCZ. During the Indian monsoon the low-level south-west monsoon winds transfer water vapour from the southern Indian Ocean and the Arabian Sea into the Indian sub-continent. At higher latitudes the warm moist poleward winds, associated with the warm sectors of extra-tropical cyclones, transfer water vapour into regions of precipitation surplus.

Figure 5.8 shows estimates for the components of the water budget over the major ocean basins and continents. Over the continents as a whole these components are much less than those over the ocean, both for evaporation and precipitation, but, on average, precipitation exceeds evaporation by about 25 cm year^{-1}. This surplus must be transferred by rivers into the ocean, where a net deficit of the same amount must exist if the balance is to be maintained. (It is noted that the net deficit over the ocean and the net surplus over land do balance when multiplied by the appropriate fractions of land and ocean areas.) The largest surplus per unit area exists over South America and the lowest over Australia. These continental surpluses must be balanced by a net flux of atmospheric water vapour from the ocean to the continents. Over the ocean it is seen that, although evaporation exceeds precipitation on average, excess evapora-tion occurs only in the Atlantic and Indian Oceans, whilst in the Pacific and Arctic Oceans precipitation exceeds evaporation. For the balance to be maintained a net transfer of fresh water from the Pacific and Arctic

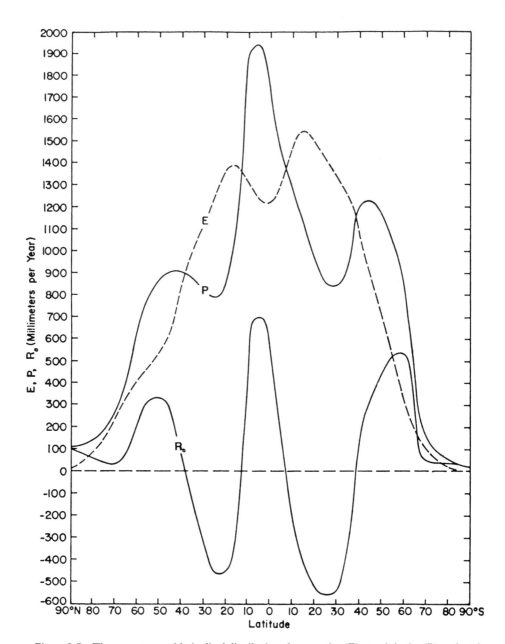

Figure 5.7. The average annual latitudinal distribution of evaporation (E), *precipitation* (P) *and total runoff* (R₀). *(Reproduced from* Physical Climatology, *by W. D. Sellers, p. 84, figure 26, with permission of University of Chicago Press.)*

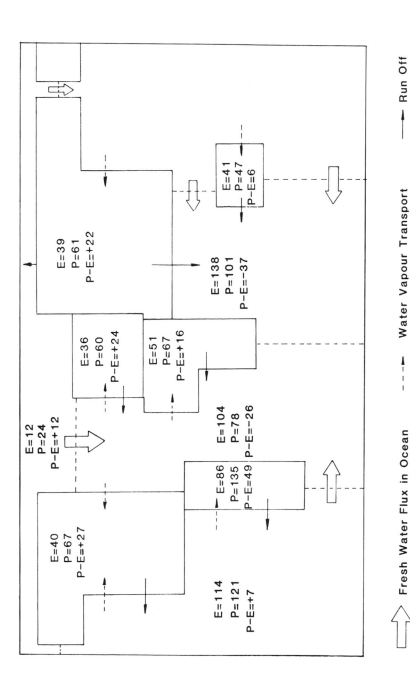

Figure 5.8. Water budget over continents and ocean basins. The broad arrows represent the freshwater flux in the ocean, the dotted arrows the water vapour transport and the solid arrows the runoff. Evaporation (E) and precipitation (P) in cm year⁻¹. (Based on data from Physical Climatology, by W. D. Sellers, p. 85, table 13, with permisson of University of Chicago Press.)

Oceans into the Atlantic and Indian Oceans is required. The water budget for the Atlantic Ocean shows that 20 cm year^{-1} of the water deficit is supplied by the surrounding continents and the equivalent of 6 cm year^{-1} comes from the Arctic and Pacific Oceans. In contrast, the Indian Ocean gains only 7 cm year^{-1} from continental runoff and the remainder of the 30 cm year^{-1} is advected from the Pacific Ocean.

In conclusion, it has been shown that the surface water budget allows a picture of the global transfers of atmospheric water vapour and of fresh water in the ocean to be ascertained. As yet, this picture is still crude, because of the inaccuracies of the precipitation and evaporation estimates, but with improved satellite techniques it should be possible to derive water budgets for seasonal and interannual variations as well as for the long-term means.

5.5. The salt budget of the ocean

The total mass of the ocean is $1 \cdot 4 \times 10^{21}$ kg and if a global average salinity of $34 \cdot 7$‰ is assumed, then the total mass of dissolved salt is $4 \cdot 8 \times 10^{19}$ kg. It has been estimated that the rate of salt input into the ocean is about 3×10^{12} kg year^{-1} and therefore the replacement time for the ocean salt is approximately 16 million years. This simple calculation illustrates the point that the sources and sinks of salt in the ocean are only of concern on a geological time scale. For ocean circulation problems, a time scale of less than 1000 years is applicable and therefore it is a good approximation to assume that the total salt content of the ocean is constant. This has the important consequence that observed salinity variations in the ocean are the result of imbalances in the water budget resulting from precipitation, evaporation and runoff.

Over the ocean basins, away from major river discharges, the salinity distribution is well correlated to the net surface water flux, W. The principal features are that the largest surface salinities occur in the sub-tropical zone, where there is a net deficit of water and that the lowest surface salinities occur in high latitudes, where precipitation exceeds evaporation. A minimum is also apparent in the region of the ITCZ. As might have been anticipated from the discussion in the previous section, there are considerable differences between the various ocean basins. The North Pacific Ocean has an average surface salinity of about $34 \cdot 2$‰ whilst the North Atlantic Ocean has a salinity of $35 \cdot 5$‰. These observations are related to the surplus of fresh water in the Pacific Ocean and the deficit in the Atlantic Ocean. Furthermore, the low salinity in the high latitudes of the North Pacific maintains a stable density stratification which inhibits the formation of deep water. The formation and stability of the halocline in the Arctic Ocean is related partly to the excess of precipitation over evaporation and partly to the influx of fresh water from runoff, particularly by Siberian rivers. A major concern has been that a reduction of runoff caused by a redirection of Siberian rivers from the Arctic Ocean to the Caspian Sea may affect the stability of the Arctic halocline. It has been argued that weakening of the halocline would enable deep convection to occur which would, in turn, bring warmer water to the surface and inhibit the development of sea ice. Although this is only a possibility, it does demonstrate the point that small variations in the water balance may produce dramatic changes in regional climate.

A useful example of the application of the principle of the conservation of salt may be found in the Red Sea. There is a surface flow of low-salinity water into the Red Sea through the Strait of Bab el Mandeb and an outflow of high-salinity water over a sill at a depth of 110 m. If F_I is the volume flux of water into the Red Sea, with a salinity of S_I, and F_O is the volume flux of water out of the Red Sea, with a salinity S_O, then for the conservation of salt across the Straits of Bab el Mandeb:

$$\varrho_I F_I S_I = \varrho_O F_O S_O \tag{5.25}$$

It can be assumed that $\varrho_I \simeq \varrho_O$ and for the conservation of water:

$$F_I = F_O - W - R_O \tag{5.26}$$

where W is the net water flux through the surface of the Red Sea and R_O is the runoff. Thus eliminating F_O:

$$F_I S_I = (F_I + W + R_O)S_O$$

$$\therefore \quad F_I\left[\frac{S_O - S_I}{S_O}\right] = -[W + R_O] \tag{5.27}$$

The salinity, S_O, is 40‰ and S_I is 37‰. Measurements across the Straits suggest $F_I = 0 \cdot 5 \times 10^6 \, m^3 \, s^{-1}$ and therefore the net surface flux of water is approximately $3 \cdot 75 \times 10^4 \, m^3 \, s^{-1}$. The negative sign indicates an excess of evaporation over precipitation and runoff. The surface area of the Red Sea is $438 \times 10^9 \, m^2$, therefore the excess evaporation over precipitation and runoff is $2 \cdot 70 \, m \, year^{-1}$. Since precipitation is generally less than 20 cm $year^{-1}$ and the runoff term is negligible, the evaporation rate is close to this figure.

Equation 5.27 is generally applicable to any semi-enclosed sea or estuary where the vertical circulation can be represented by a two-layer fluid (figure 5.9). Some conclusions can be drawn from this equation. When evaporation exceeds precipitation and runoff, $(W + R_O) < 0$ and

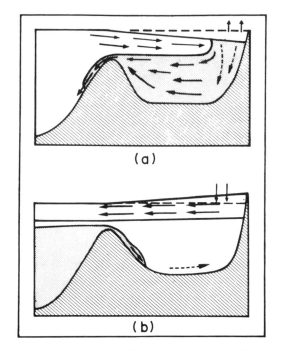

Figure 5.9. The two principal types of sill basin. (a) Evaporation > precipitation + runoff. (b) Evaporation < precipitation + runoff. (Reproduced from Descriptive Regional Ocean- ography, *by P. Tchernia, p. 38, figure 3.5, with permission of Pergamon Press.)*

$S_O > S_I$, i.e. the outflow will be more saline than the inflow. If the temperature changes are much smaller than the salinity changes, then the more-saline water will sink and flow out underneath the less-saline inflow. This situation is typical of the Red Sea, the Persian Gulf and the Mediterranean Sea, all of which produce saline bottom waters. In the northern extremities of the Red Sea and in the Persian Gulf salinities as high as 42‰ occur, whilst in the Mediterranean Sea salinities of 38‰ are typical. In higher latitudes, precipitation and runoff exceed evaporation, i.e. $W + R_O > 0$, and therefore $S_O < S_I$. Outflow will thus be less saline than inflow and this fresher water will flow out of the surface layer whilst the more-saline ocean water will flow in at depth. This situation is representative of the Baltic Sea, where surface salinities as low as 10‰ can occur. From equation 5.27 it can also be seen that the salinity difference, $S_O - S_I$, is proportional to the ratio of $(W + R_O)/F_I$ and, if the volume flux into the basin is considerably larger than the net water flux, $W + R_O$, then the salinity difference between the inflow and outflow will be small. On the other hand, large salinity changes will occur if the volume flux is small relative to the net water flux. The Baltic Sea, for instance, has a very shallow sill depth of 18 m, the volume flux is very small compared with the freshwater input and hence the surface water has a low salinity.

The value of equation 5.27 lies in obtaining either a consistency check on the salt budget of an estuary or semi-enclosed sea or in determining the value of one of the terms in the equation, such as precipitation which, as previously discussed, is difficult to measure directly. The principle can also be applied to the salt balances of ocean basins such as the Arctic Ocean. In this case, the situation is complicated by the large number of inflows and outflows and by the fact that the vertical circulation may be rather more complicated than that assumed for the Red Sea. In ocean basins such as the Atlantic Ocean direct measurements of the volume fluxes of the water masses are only possible in limited areas, but equations based on the conservation of salt, mass and heat have been solved, using inverse analysis, to provide large-scale transports.

5.6. *Temperature and salinity relations in the ocean*

In the previous section it was shown how the surface water flux, runoff and circulation determine the observed salinity and it is reasonable to acknowledge that the surface heat budget, in conjunction with circulation, can determine the observed temperature of the ocean. It is clear that the relationship between temperature and salinity will give useful information about the sources of water masses and about the circulation and mixing rates in the ocean. Close to the surface the temperature and salinity characteristics will be constantly changing in response to synoptic and seasonal variations in heat fluxes and water fluxes. However, below the depth of seasonal influence, on average about 100 m, the temperature and salinity of a water parcel will only change by mixing with the surrounding water masses and therefore it can be said that the temperature–salinity (T–S) properties of the parcel will be conserved. In the deep ocean, potential temperature, θ, is used rather than the actual temperature, T, because of the adiabatic compression.

Figure 5.10 shows the temperature and salinity profiles taken at two oceanographic stations in the Gulf Stream, about 200 km apart. At first glance there is no obvious similarity but, when the information is plotted on a T–S diagram, also shown in figure 5.10, it can be seen that they have an identical characteristic shape. In this case, the water mass at one station has been vertically displaced by 600 m relative to the other station. This example shows that a T–S relationship will allow a particular water mass to be labelled and to be followed in the ocean. In oceanography a point on a T–S diagram is referred to as a water type whilst a straight-line relationship is known as a water mass. Thus, the water type refers to the

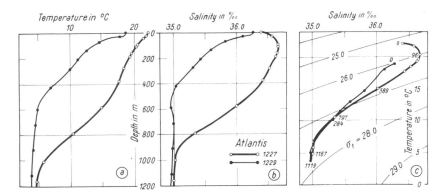

Figure 5.10. Relation between temperature and salinity for two neighbouring 'Atlantis' stations, 1227 and 1229, in the Gulf Stream east of Cape Hatteras. (a) Vertical temperature distribution. (b) Vertical salinity distribution. (c) T–S diagram. (Reproduced from General Oceanography, *by G. Dietrich, p. 197, figure 78, with permission of Interscience.)*

original T–S during formation at the surface whilst the water mass evolves by mixing between different water types. Figure 5.11 shows the progressive mixing of two water types, A and B. In order to conserve temperature and salinity, mixtures of the two water types must lie on the line *AB*, and the position on the straight line is determined by the relative concentration of each water type. Hence a succession of different water types are formed. The ensemble of water types is known as a water mass. An example of the progressive mixing of a water type is the distribution of the Antarctic intermediate water in the Atlantic Ocean, as shown in figure 5.12. This water type is found in the Antarctic convergence zone at about $50°S$ where it has a temperature of $2 \cdot 2°C$ and a salinity of $33 \cdot 8$‰. It sinks to a depth between 700 and 1500 m, and then it progresses northwards as a tongue of low-salinity water. It can be identified on the T–S diagram of the Atlantic Ocean, shown in figure 5.12, as a salinity minimum. Successive T–S diagrams show that the water mass becomes warmer and more saline as it mixes with the central Atlantic water above and the Atlantic deep-water below. Eventually, at about $20°N$, it can no longer be identified on a T–S diagram. From the T–S diagrams the rate of mixing of the Antarctic intermediate water can be deduced and thus an estimate of the volume of the original water type can be made. Figure 5.11 shows the mixing of three water types, not dissimilar to those in the Atlantic Ocean. The mixed water will be located within the triangle *ABC* and its actual position will depend on the relative proportions of the three types. If the rate of mixing between the water mass above and the water mass below is equal, then the salinity minimum will be located along the line *BD*. With the Antarctic intermediate water

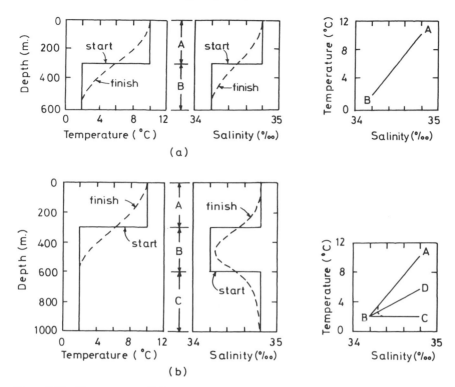

Figure 5.11. Temperature–salinity relationships resulting in the mixing of water masses. (a) Mixing of two water types. On the T–S diagram, mixing of the two water types is shown by the straight line connecting the T–S points characteristic of the two original water types. (b) Mixing of three water types. On the T–S diagram, the mixing process is shown by two straight lines indicating that water types A and B mix, and that water types B and C mix. The lines meet at a point representing water type B, now altered by mixing with water above and below so that its original identity has been lost and it can no longer be represented by a single point on the T–S diagram (---). For equal mixing above and below, the salinity minimum associated with water type B will lie along the line BD. (Reproduced from Oceanography: A View of the Earth, *by M. Grant Gross, p. 449, figure A4-1, with permission of Prentice-Hall.)*

it is indeed the case that the mixing is approximately equal between the central Atlantic water and the Atlantic deep water.

Figure 5.13 shows a three-dimensional diagram representing a census of the world's water masses by Worthington (1981). The vertical co-ordinate is the volume of water having a mean temperature in the range $\mp 0 \cdot 05°C$ and a mean salinity in the range $\mp 0 \cdot 005‰$. Thus, the peak in figure 5.13 corresponds to a potential temperature between $1 \cdot 1$ and $1 \cdot 2°C$ and a salinity between $34 \cdot 68$ and $34 \cdot 69‰$ and has a volume of $26 \times 10^{15} \text{ m}^3$. This water mass has a larger volume than all of the water above $19°C$ and larger than the total volume of all the continental ice caps. The census shows that the major water masses of the ocean have a

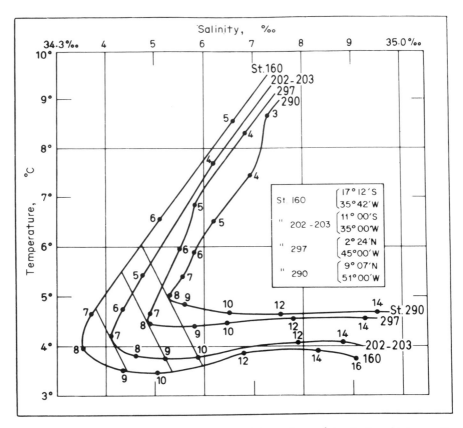

Figure 5.12. Variations with latitude of the salinity minimum characterizing the Antarctic intermediate water between 17°S and 9°N in the Atlantic. (Reproduced from Descriptive Regional Oceanography, *by P. Tchernia, Pergamon Marine Series Vol. 3, p. 157, figure 5.34, with permission of Pergamon Press.)*

relatively small range of temperature and salinity, distributed about a global mean temperature of $3 \cdot 51°C$ and a mean salinity of $34 \cdot 72‰$. The much warmer surface waters and the saline deep waters of the Mediterranean and Red Seas are insignificant compared with the deep polar waters which fill all of the ocean basins. Thus the major water masses of the ocean are probably not significantly different to those occurring during the Ice Ages and the amelioration of the climate by the ocean is limited to the effect of a rather shallow, warm surface layer. The most uniform water masses are found in the North Pacific Ocean where there are no sources of bottom water. All these deep-water masses have their origin in the Atlantic Ocean and therefore the mixing process has had sufficient time to homogenize the deep-water masses in the Pacific. The Indian Ocean tends to have a relatively uniform salinity of $34 \cdot 70‰$

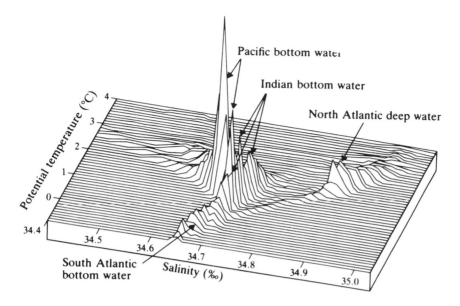

Figure 5.13. *Simulated three-dimensional T–S diagram of the water masses of the world ocean. (Reproduced from L. V. Worthington, chapter 2, p. 47, figure 2.2, in* Evolution of Physical Oceanography, *ed. by B. A. Warren and C. Wunsch, with permission of MIT Press.)*

but a larger range of temperatures. The Atlantic Ocean is quite different from either the Indian or the Pacific Oceans. It has a large number of water masses ranging from -1 to $+4°C$ and most are relatively saline compared with the Indian and Pacific Oceans. The large spread of water masses is indicative of incomplete mixing which is in agreement with the hypothesis that the Atlantic Ocean is the source region for much of the deep water. Of note is the formation of North Atlantic deep water in the Labrador and Greenland Seas during late winter and the formation of Weddell Sea bottom water which, at $27 \cdot 88 \, \sigma_t$, is the densest water mass in the ocean, having a temperature of $-0 \cdot 5°C$ and salinity of $34 \cdot 66‰$. Dense water is also found at the bottom of the Norwegian Sea, the Red Sea and the Mediterranean Sea but, because of the sills at the entrances to these seas, these dense waters cannot flow freely into the open ocean.

A novel application of the T–S relationship has been made recently by Stommel and Csanady (1980) to estimate the ratio of the freshwater flux to the heat flux. By analogy with equation 5.5, the gradient of salinity is proportional to the freshwater flux and the gradient of temperature to the heat flux. By taking the ratio of the temperature gradient to the salinity gradient, Stommel and Csanady obtained the ratio of the freshwater flux to the heat flux.

Most of the T–S information which appears in the scientific literature was collected over the last century by research cruises. The classical method of measuring temperature at depth is by using two accurate reversing thermometers. One thermometer is protected from adiabatic compression by enclosure in a strong glass tube whilst the second thermometer remains unprotected. The difference in the temperatures measured by the two thermometers is, therefore, indicative of the hydrostatic pressure of the water from which the depth can be determined to an accuracy of ∓ 20 m. The reversing thermometer, as the name suggests, works on the principle that rotating the thermometer vertically by $180°$ will break the column of mercury between the capillary and the bulb, and the thermometer will record the temperature of the water at the depth of the reversal. The thermometers can be subsequently recovered from deep water to give a determination, accurate to $\mp 0\cdot01°C$, of the *in situ* water temperature. In practice, pairs of thermometers are attached to water samplers, or Nansen bottles, which are placed at regular intervals along a wire cable. The cable is subsequently winched down into the water column. Once the Nansen bottles are at the required depth a small weight, known as a messenger, is sent down the cable. The messenger trips the top clamp of the Nansen bottle which is then reversed together with the attached thermometers. The reversal of the bottle traps a sample of the *in situ* water which can be recovered and analysed for salinity in the ship's laboratory, either by titration or by conductivity measurements. After reversal, a second messenger is sent down the wire to release the next Nansen bottle and so on. This method of T–S measurement is still widely used today but it does have a major limitation in the vertical resolution obtained. Ten Nansen bottles are typically used for each cast and in waters of, say, 4000 m depth the sampling frequency would be one T–S measurement for every 400 m. It may be possible to do repeated casts at the same station, but this is very time consuming as each cast can take between 1 and 2 hours.

A considerable improvement in vertical resolution and precision has been obtained from the development of the salinity, temperature and depth recording system (STD) and conductivity, temperature and depth recording system (CTD). The STD and CTD systems consist of three sensors whose resistances vary with the temperature, conductivity and pressure of their surroundings. The resistances of the three sensors are measured by an audio-oscillator whose frequency is determined by the resistance. The signals are sent up a cable to be processed on board the research ship. The conductivity of the sea water is measured using platinum electrodes and it is dependent on both temperature and salinity. Thus a temperature correction to the conductivity signal is required to obtain accurate salinity values. Usually two temperature sensors are employed. One is a thermistor, which has a very fast response time, and the other is a platinum resistance thermometer, which is very accurate. The thermistor measures small variations in temperature and the platinum resistance thermometer is used in the absolute calibration of the thermistor and the conductivity sensor. By this method, *in situ* temperatures can be measured to $\mp 0 \cdot 001^\circ$C, salinity to $\mp 0 \cdot 005$‰ and depth to an accuracy of 1 m. Additional probes can be attached to measure sound velocity and dissolved oxygen in the water column. Thus, from one cast of a CTD system, a high-resolution vertical distribution of temperature and salinity can be obtained.

These new measurements have shown that the T–S relationships for water masses are not as uniform as shown by classical methods. Distinct layers of uniform water have been found to be separated by rapid changes in temperature and salinity. These uniform layers range from as little as 1 m in depth up to 100 m. The small-scale variations in T–S relationships are known as the fine structure and it is now believed that these fine structures are direct evidence for water-mass transformation in the deep ocean. One process which can cause this fine structure at the boundaries of water masses is double-diffusive convection. One example is salt fingering, which depends on the much larger diffusivity of heat compared with the molecular diffusivity of salt. Consider two layers of water, the upper one being salty and warm whilst the lower layer is cold and fresh. If it is assumed that temperature effects dominate the density of each layer, the two fluids will be statically stable. Consider now a parcel of fluid moved from the upper layer to the lower layer where, because of the larger heat diffusivity, it will lose heat more rapidly than it will lose salt. The parcel will lose its buoyancy, will become statically unstable and sink. A parcel moving from the lower layer to the upper layer will gain heat more rapidly than salt, will become buoyant with respect to its surroundings and accelerate upwards. Hence alternating upward (fresh) and downward (salty) motion is produced, which transfers salt downwards across the interface. The release of convective energy by the salt fingers results in the formation of well mixed layers above and below the interface. Such fine structure has been found under the Mediterranean outflow where salty and relatively warm water from the Mediterranean Sea overlies the less saline, but colder, North Atlantic deep water.

A second process, which can lead to mixing between water masses, depends on the non-linear equation of state for sea water and it is referred to as 'caballing' in the oceanographic literature. Consider two parcels of fluid having different temperatures and salinities but equal densities, as depicted in figure 5.14. The resultant mixture will lie along the line *AB*, its position being determined by the relative proportions of the two original fluids. However, because of the curvature of the lines of constant density known as isopycnals, the mixture will be more dense than the original water parcels and therefore the mixture will sink. The curvature

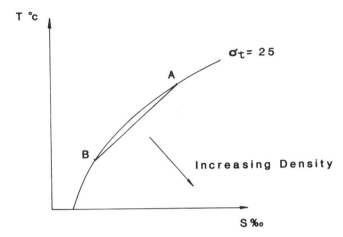

Figure 5.14. Mixing of two water masses (A, B) having similar densities ($\sigma_t = 25$). Any mixture always has a density $\sigma_t \geqslant 25$.

of the isopycnals is a maximum at low temperatures and caballing is, therefore, particularly important in Antarctic and Arctic waters. The formation of Antarctic bottom water has been attributed to this process. In the Weddell Sea the surface water is generally colder and less saline than the underlying Antarctic circumpolar water. In winter, the formation of sea ice releases extra brine into the surface layer which, in conjunction with winter cooling, may produce a water mass of similar density to the underlying one. In this case mixing of the upper and lower water masses may produce a mixture which has a density greater than the deep water and which will therefore sink.

The above fine-structure processes are not the only ones by which water masses are diluted by mixing. In the case of outflows of dense water from semi-enclosed seas, the plumes may be rapidly mixed by entrainment with the surrounding water. Occasional intense gravity currents are observed over the Faeroes–Iceland ridge and in the Faeroes Channel, when dense Norwegian Sea water flows over the sills into the Atlantic Ocean. The plume accelerates down the slope because of the density differences with the ambient water, and may reach speeds of 1 m s^{-1}. Intense mixing of the plume occurs, which increases its volume of water and reduces its density, to form North Atlantic deep water. Calculations of the mixing ratio have shown that the overflow water is composed of $1 \times 10^6 \text{ m}^3 \text{ s}^{-1}$ Norwegian Sea water and $3 \times 10^6 \text{ m}^3 \text{ s}^{-1}$ entrained Atlantic water. The dense high-salinity plume from the Red Sea also experiences mixing with Indian Ocean water as it moves off the Bab el Mandeb sill into the Gulf of Aden. Here, the plume descends to a depth of 1000 m before it reaches a similar density to the surrounding water and spreads horizontally into the Indian Ocean.

A second important horizontal mixing process is caused by energetic mesoscale eddies, typically 50–200 km in diameter. These eddies have been found in all parts of the ocean and can homogenize water masses, especially in intense frontal zones such as the Gulf Stream and the Antarctic convergence zones. It is thought that the diffusion of Mediterranean water into the North Atlantic is primarily due to horizontal mixing by eddies. In 1978 a water mass having a temperature and salinity characteristic of Mediterranean water was located in the western Atlantic in a small intense mesoscale circulation. It had apparently taken about 3 years to cross the Atlantic Ocean and in that time it had not mixed, to any significant extent, with its environment. This observation suggests that there is still much to be understood about the mixing of water masses in the ocean.

5.7. Tracers in the ocean

In the early 1970s a research programme known as GEOSECS (Geochemical Ocean Sections Study) was initiated to measure the distribution of natural and anthropogenic tracers in the world ocean. Although oceanographers had used naturally occurring tracers, such as dissolved oxygen and carbon-14, to study water masses before this experiment, GEOSECS was the first comprehensive attempt to measure these tracers, and many others, on a global basis.

Table 5.2. shows tracers which have been successfully used in the ocean to determine water-mass production, mixing and circulation. Dissolved oxygen has been measured in the ocean, in conjunction with salinity and temperature, since the 1930s and much information on its global distribution is available. Oxygen is absorbed most efficiently at low temperatures and thus the largest concentrations are found in Antarctic and Arctic waters. It is a good tracer for studying the mixing of water masses, such as the North Atlantic deep water and Antarctic bottom water, but it has to be monitored continuously because it is consumed by bacteria and animals even in the deep ocean. Nitrates and phosphates are excreted by animals and bacteria and they have been used as short-term tracers. Recently it has been realized that the total sums of dissolved oxygen and phosphate, and dissolved oxygen and nitrate, tend to be quasi-conservative tracers even in the presence of living organisms, although they cannot be used in waters with oxygen concentrations of less than 1%.

Silica is produced from siliceous bottom sediments such as red clays,

Table 5.2. Selection of artificial and natural tracers in the ocean.

Tracer	Origin	Conservative	Half-live years	Comment
Dissolved O_2	Surface exchange	No	—	Consumed by bacteria and zooplankton
Silica (SiO_2)	Bottom sediments	No	—	Removed by upper ocean
Nitrate (NO_3)	Bacteria and zooplankton	No	—	—
Phosphate (PO_4)	Bacteria and zooplankton	No	—	—
$[O_2] + 9[NO_3]$	—	Quasi-conservative	—	Not conserved in low O_2 water
$[O_2] + 135[PO_4]$	—	Quasi-conservative	—	—
^3H	Cosmic rays Nuclear testing	Conservative	12·4	Natural tritium very small compared with bomb tritium
^{90}Sr	Nuclear testing	Conservative	28·6	Concentrated in tropical waters Recorded in growth rings of coral
^{14}C	Cosmic rays Nuclear testing	Conservative	5730	Appears in solution and particulates
^{137}Cs	Nuclear testing	Conservative	30·2	Require large samples for measurement
^{39}Ar	Cosmic	Conservative	250	Require very large samples

which are plentiful in the Pacific Ocean and, to a lesser extent, in the Indian and Atlantic Oceans. Silica is consumed by biological processes in the upper layers of the ocean and it is, therefore, most useful for studying bottom and intermediate water circulations. In the North Atlantic deep water, the silica concentrations are low compared with those in the Antarctic circumpolar waters and Antarctic bottom waters and it is, therefore, an ideal tracer when studying the mixing of North Atlantic deep water into the Antarctic Circumpolar Current. North Atlantic deep water is composed of a number of different water types and so the θ–S characteristics vary accordingly. However, the silica concentration is remarkably uniform and it is therefore a more consistent tracer than the θ–S profile. Figure 5.15 shows the North Atlantic deep water (low silica concentrations) flowing south against the western boundary, and the Antarctic bottom water (high silica concentration) flowing northwards along the bottom. It is noted that in bottom waters, where the θ–S properties usually show only a small variation, the use of silica as a tracer is particularly useful.

The information obtained from the above natural tracers is only qualitative, because of the lack of knowledge about rates of production and consumption. Quantitative information on the age and circulation time scale of any given water mass can be obtained using radioactive tracers. As table 5.2 shows, some tracers are produced naturally from cosmic rays, such as ^{39}Ar, some are by-products of nuclear tests, such as ^{90}Sr and ^{137}Cs, and others, such as tritium, ^{3}H and ^{14}C, are by-products of both nuclear tests and cosmic rays. The long-lived, naturally produced isotopes are particularly useful because they are generally in a steady-state balance in the ocean and therefore they can be used to estimate the age of the ocean waters and the ventilation time of the ocean. ^{14}C measurements, obtained before nuclear bomb tests significantly altered the ^{14}C distribution in the ocean, enabled the first estimate of the ventilation time, T, to be obtained thus:

$$T = V/R_v$$

where V is the volume of the ocean and R_v is the ventilation rate for deep water. For radiocarbon the ventilation time is also given by:

$$T = \left[\frac{(^{14}C/C)_S}{(^{14}C/C)_D} - 1 \right] T(^{14}C) \qquad (5.28)$$

where $(^{14}C/C)_S$ and $(^{14}C/C)_D$ are the carbon isotopes ratios for the surface and deep waters, respectively, and $T(^{14}C)$ is the mean half life of ^{14}C, which is 8200 years.

Equation 5.28 gives a mean ventilation time of 1400 years $\mp$ 250 years

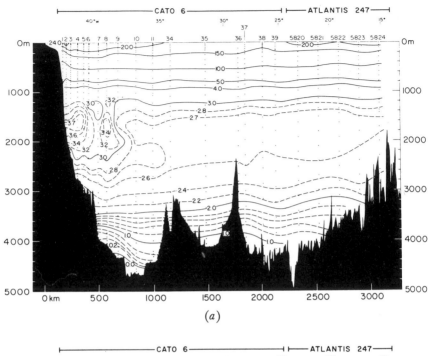

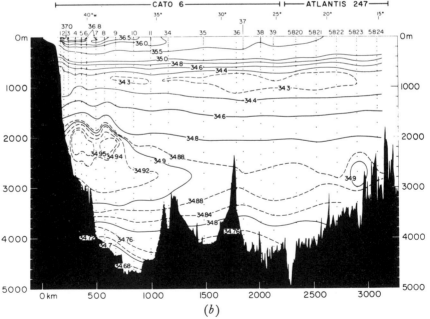

Figure 5.15. Sections of (a) potential temperature (°C), (b) salinity (°/oo), (c) concentrations of dissolved oxygen (ml l⁻¹) and (d) silica (ml l⁻¹) along roughly latitude 30°S from South America (left) to the Mid-Atlantic Ridge, illustrating the two deep western boundary currents of the South Atlantic,

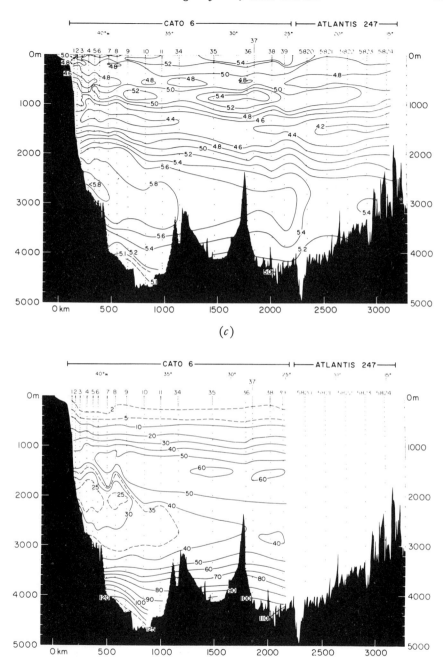

namely, the northward flowing Antarctic Bottom Water and the southward flowing North Atlantic Deep Water above. (Reproduced from J. L. Reid et al., J.P.O., **7**, 62–91, figure 3a and 3b, with permission of the American Meteorological Society.)

for deep-water circulation. Recent evidenc⌄ has suggested that this is an overestimate, because it is often not possible to distinguish old water masses which may have been recirculated around the ocean a number of times from newly formed water masses. Measurements on North Atlantic deep water have indicated an apparent age of between 80 and 225 years. Argon is also a potentially useful natural radioactive tracer because it has a half-life similar to the ventilation time. However, at present, it is technically difficult to determine because several cubic metres are required to obtain just one measurement.

The radioactive tracers produced from atmospheric nuclear-weapon tests between 1945 and 1963 have inadvertently provided a number of useful transient tracers. These tracers are not in equilibrium with the ocean and they cannot provide information on the age of water masses, but they have proved valuable for determining the progress of recently formed bottom and intermediate waters. The bomb isotopes best studied in the GEOSECS experiment were ^{14}C and ^{3}H. Anthropogenic carbon-14 can be distinguished from natural carbon-14 because of its very large signal in the surface layer. The distribution of tritium is not known prior to the bomb tests but it is believed that the natural concentrations are much lower than the anthropogenic ones. ^{14}C and ^{3}H were released

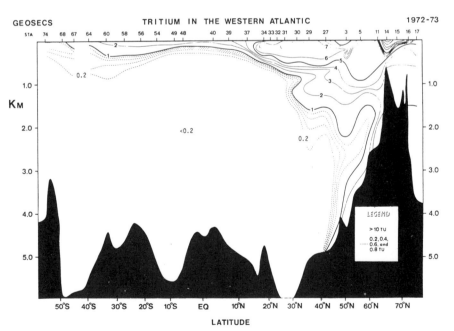

Figure 5.16. Tritium section along the western basin of the Atlantic Ocean obtained during the GEOSECS programme. (Reproduced from Ostlund and Fine (1979), I.A.E.A. –SM–232167, with permission of the IAEA.)

into the Northern Hemisphere's stratosphere, leaked into the troposphere about 1 year later and entered the ocean by precipitation. By 1973 about 45% of the total bomb ^{14}C had entered the ocean and penetrated to an average depth of 300 m. In the surface layers, the largest values of both isotopes are located in the Pacific and Atlantic Oceans at about 50°N. The concentrations decrease towards the Southern Hemisphere but the concentration gradient is not uniform. Near 15°N there is a remarkable reduction in both ^{14}C and ^{3}H concentrations, as shown in figure 5.16, which is associated with a frontal boundary between the sub-tropical gyre to the north and the equatorial waters to the south. The low concentrations and limited vertical penetration between 15°N and 15°S suggest that water is being upwelled from depths of at least 500 m. Broecker (1978) has shown that this equatorial upwelling flux of $45 \times 10^{6} \, \mathrm{m}^{3} \, \mathrm{s}^{-1}$ is comparable with the total production of bottom water in the global ocean. The most remarkable and unexpected finding, shown in figure 5.16, is the relatively high concentrations of ^{3}H in the North Atlantic deep water only 20 years after its surface input, thus indicating a more rapid ventilation of the deeper ocean than had previously been anticipated. In the western Atlantic, the southward bottom boundary current has been identified by its high tritium concentrations.

It is expected that future monitoring of both ^{3}H and ^{14}C in the ocean will reveal more about the movement of intermediate- and deep-water masses as this inadvertent experiment continues.

6

Observations of winds and currents

6.1. Measurement of winds and currents

Both atmosphere and ocean are fluid environments and the basic techniques of the measurement of winds and currents have close similarities. There are two fundamentally distinct ways of specifying a flow field in the atmosphere and the ocean. One approach, known as the Eulerian description, expresses the flow field at a fixed position in space relative to the Earth's surface. For example, a meteorologist may obtain an Eulerian description of the surface wind field from measurements of the wind direction and speed at an array of weather stations at a particular instant in time. Mathematically, an Eulerian velocity flow can be defined as a vector quantity at a position x, y, z, and at a time t:

$$\mathbf{v} = \mathbf{v}(x, y, z, t) \tag{6.1}$$

The co-ordinates in both fluids refer to an eastward direction, x, a northward direction, y, and vertically upwards, z.

The second approach, referred to as a Lagrangian description, measures the position of the chosen element of fluid as a function of time. This method gives information on the time history of a parcel of air or water. The flow is defined by the vector quantity:

$$\mathbf{v} = \mathbf{v}(\mathbf{a}, t) \tag{6.2}$$

where $\mathbf{a}$ is the position vector of the chosen element of fluid with respect to a defined origin. An example of the Lagrangian specification is the trajectory of a balloon moving with the velocity of the air around it.

It is important to note that the two specifications of the flow field are independent and therefore they give different types of information on atmospheric and oceanic flows. It is not possible to obtain *exact Eulerian* velocities from the trajectories of balloons or floats and, similarly, it is not possible to determine *exact Lagrangian* velocities from Eulerian measurements.

However, as will be shown presently, it is possible to make approximate comparisons between the two systems of measurement if sufficient simultaneous measurements of Lagrangian and Eulerian velocities are available.

The choice of measuring system depends on the type of information required. A Lagrangian measuring system using, say, drifting ocean buoys would be appropriate for determining the movement and dispersal of a natural tracer or a pollutant. On the other hand, a weather forecaster prefers the Eulerian system for specifying winds since he is most concerned with prediction over a specific area. As we have seen above, the distribution of fixed synoptic stations essentially gives an Eulerian description of the circulation and it is not surprising that numerical weather-prediction models have been designed on an Eulerian grid system for the solution of the equations of motion.

It is important to note the different conventions for specifying the flow direction in the atmosphere and the ocean. Wind is usually defined by the direction from which it has come whilst current directions are defined by the direction in which they are going. Hence a westerly wind is flowing from the west to the east, and this would correspond to an eastward current in the ocean. From hereon, the oceanographic convention for both wind and currents is adopted wherever possible.

In both the atmosphere and the ocean, Eulerian velocities are most commonly measured by a rotary device, such as an anemometer or current meter, where the number of rotations of a set of wings or propellers in a given time is measured. The anemometer has to be secured from a tower or mast and therefore it can only be used for wind measurements near the Earth's surface. Most cup-anemometers are sensitive to wind variations longer than 10 s and have an accuracy of 1 m s^{-1}. Smaller and lighter anemometers can be used for more accurate observations of the wind close to the ground, such as required in micrometeorologial experiments. The main difficulty with anemometer measurements is the problem of the alteration of the flow by obstructions, in particular by the tower or mast. For example, a ship may distort the air flow and this will lead to errors in wind measurement which are dependent on the orientation of the ship to the wind direction and the exposure of the anemometer.

The current meter is now the mainstay of modern Eulerian measurements in the ocean. In recent years considerable improvements in the technical design of current meters have been made, coupled with better mooring and data systems. A commonly used meter is the Savonius meter which consists of a set of cups which rotate in proportion to the flow through the meter. The direction of flow is obtained from a freely moving direction vane attached in line with the instrument. The Savonius meter can measure current velocities to an accuracy of 2 cm s^{-1}. Though the original instrument was designed in the 1930s, it has been improved upon by using a magnetic-tape storage system to record velocity and direction at about eight times per rotor revolution. The instrument samples bursts of data every 15 min because of the high-frequency variability of the currents. It automatically vector averages bursts of data to give an accurate measurement. This is important close to the surface where waves produce low-frequency noise. One unfortunate problem with the Savonius rotor is its susceptibility to vertical motion induced by surface waves and this alone is capable of inducing the rotor to move. The problem does not arise with the vector-measuring current meter which uses two horizontal axis propellers orientated in perpendicular directions. This automatically measures both horizontal components of the flow and it is unaffected by surface wave motion. In all current meters, the orientation of the instrument is measured by a magnetic compass.

A major difficulty in the Eulerian measurement of currents at sea is the provision of a stable platform at a known position. Current measurements from ships are subject to the ship's motion and care must be taken in their interpretation. Most current measurements are now taken from surface and bottom moorings which are laid by an oceanographic vessel and left in position for up to 2 years before being recovered. Typical types of mooring are shown in figure 6.1. The main features of the moorings are a radio, a light and a float, all necessary to locate the mooring at the end of an experiment. Current meters, and other instruments, are suspended at the required intervals down to the ocean bottom. Moorings usually have between five and ten current meters attached to the wire or the nylon cable. Buoyancy is necessary because of the weight of the instruments and cable and it is provided by glass spheres encased in plastic containers. Finally the wire is anchored to the bottom. A feature of some moorings is the acoustic release system which detaches the cable from the bottom anchor when it is activated by a research ship at the end of the experiment. In the early days of development the recovery rate was about 50% but now, with the advent of more robust systems, the recovery rate for moorings is about 95%. Some of the problems that have had to be overcome were from fish severing the mooring lines, from corrosion of the mooring wires and from intense surface currents pulling the floats under. Surface moorings have a higher failure rate than sub-surface moorings because of the considerable stresses induced by surface wave motion. Although moorings have proved a cost-effective way of obtaining Eulerian current observations, the movement of the mooring, particularly at or close to the surface, produces errors. This type of motion can be measured by acoustic tracking of the mooring and it is therefore possible to compensate for it when analysing the current measurements.

Other Eulerian techniques include the determination of the angle of tilt of a wire supporting a known drag. This is useful for measuring current

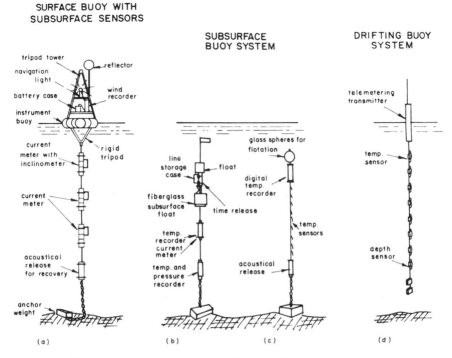

SURFACE BUOY WITH
SUBSURFACE SENSORS

SUBSURFACE
BUOY SYSTEM

DRIFTING BUOY
SYSTEM

Figure 6.1. Schemes for mooring current meters. (a) Surface buoy with meteorological instruments and sub-surface current meter. (b) Sub-surface buoy with water property and current measuring instruments. (c) 'Chain' of temperature sensors with recorder. (d) Drifting temperature chain. (Reproduced from Descriptive Physical Oceanography, *fourth edition, by G. L. Pickard and W. J. Emery, p. 88, figure 6.3, with permission of Pergammon Press.)*

velocities near the surface but, because the drag force increases with depth, it can give misleading determinations of the currents at depth.

Sound measurements can be used to determine Eulerian velocities by two different methods. The first method requires the measurement of the travel time of sound between two hydrophones A and B, as shown in figure 6.2. The travel time between A and B is $L/(c+v)$, where v is the average current velocity, c is the speed of sound and L is the distance between A and B. The travel time between B and A is $L/(c-v)$. In an acoustic current meter L is known and thus v may be determined, without knowledge of the velocity of sound c, from the difference between travel times. Of course, the system as shown will only measure the current velocity in the direction of the sound waves and so a second pair of hydrophones is required in the perpendicular direction to measure both components of the horizontal flow. Munk and Wunsch (1983) have

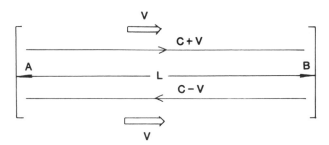

Figure 6.2. *Principle of the acoustic current meter. c is the velocity of sound and v is the water velocity.*

suggested that this technique be used to measure horizontally averaged velocities on ocean-basin scales by using an array of moored transceivers in the SOFAR channel at a depth of about 1000 m. Such an application would not, strictly speaking, yield truly Eulerian measurements.

The second acoustic method consists of two narrow-beam hydrophones which are focused on a common patch of water. The sound waves are back-scattered to the hydrophones by particles in the water and the Doppler shift in frequency between the emitted and back-scattered radiation, ΔF, is measured. The velocity of the patch of water relative to the hydrophones is obtained from the following relationship:

$$v = \frac{c\,\Delta F}{F_0 \cos \theta} \tag{6.3}$$

where F_0 is the frequency of the emitted sound waves and θ is the angle between the two hydrophones. The dependence of the velocity of sound, c, on temperature and salinity implies that it is only possible to use the Doppler method where the temperature and salinity gradients over the measurement path are small.

A useful technique in regions of high current velocity is the measurement of the emf induced in a cable moving through the Earth's magnetic field because of the action of a current. The emf, e, is related to the velocity, v, by

$$e = vlH \sin \theta \tag{6.4}$$

where θ is the angle subtended between the current and the Earth's magnetic field, H, and l is the length of the cable. This method has been successfully applied to the flow of the Florida Current and to flow through the Dover Straits. Unfortunately, only one component of the velocity is measured and the technique's main use is in monitoring unidirectional confined currents.

A similar technique, known as the Geomagnetic Electrokinetograph

(GEK) method, can be used from a ship, measuring the induced emf between two electrodes. To measure the two components of the flow, the ship is first directed $45°$ west and then $45°$ east of its original heading. This technique is not widely used but it is a valid one in regions of high current velocities.

In the atmosphere Eulerian methods include the pressure anemograph. This instrument measures the difference between the dynamic pressure facing the wind and the static pressure. The pressure difference is

$$\tfrac{1}{2}\varrho_A v^2$$

where ϱ_A is the density of the air and v is the wind velocity. This technique is very accurate and it is sensitive to rapid fluctuations in the wind. It is able to measure fluctuations of up to 1000 Hz and is particularly useful for turbulence measurements. It can be used, *in situ*, on a mast or tower, or from an aircraft. Horizontal components of the flow can be obtained by using two instruments in perpendicular directions. For very accurate measurements of small-scale turbulence, especially in the boundary layer, a number of techniques are applicable. These include the hot wire anemometer, which is very sensitive to wind fluctuations, and sonic anemometers. However, it is important to note that, for synoptic data acquisition for weather forecasting, truly Eulerian techniques are limited to surface observations.

Above the Earth's surface, horizontal wind measurement depends on the use of Lagrangian techniques; the most widely used being the tracking of a balloon by radar, theodolite or satellite. During the Second World War, when balloon tracking became important for obtaining measurements of the winds in the upper troposphere for military aircraft operations, small pilot balloons of known positive buoyancy were released from weather stations and tracked by theodolite. The observer could calculate the ascent rate from the known buoyancy of the balloon at the surface and by accurate timing could, therefore, obtain the x, y, z co-ordinates of the balloon at any time. This type of operation could be carried out by single, skilled observers, but it was limited by the height of the cloud base and the assumption of a constant ascent rate. Radar tracking systems permit the range, elevation and azimuth to be obtained automatically and thus the exact trajectory of the balloon can be traced in all weathers throughout the entire depth of the troposphere and the lower stratosphere. These balloon ascents are not, by definition, exact Lagrangian measurements because they are sampling different parcels of air as they move upwards through the atmosphere. Furthermore, because they are taking a profile of the atmosphere, the balloon may not be moving exactly with the horizontal velocity of the air at any particular level. However, despite these problems, they provide a valuable source of information on the upper-level winds for weather forecasting purposes. The winds are obtained by taking differences in horizontal positions at, for example, 1 min intervals. A balloon may take about an hour to attain a height of 20 km and during that time it may travel a horizontal distance of 50–100 km, depending on the strength of the upper winds. Meteorologists involved in weather prediction are mainly interested in mapping synoptic motions with scales greater than 1000 km. By comparison the horizontal distances travelled by the balloons are small and the measured winds are, therefore, a reasonable approximation to the Eulerian winds at the weather station.

More truly Lagrangian measurements are obtained from constant-level balloons which are designed to float at a specified pressure level in the atmosphere by appropriate adjustment of the buoyancy. These balloons are much larger than their synoptic counterparts and individual balloons have been tracked around the globe using satellite position-fixing systems. In the 1970s a large number of these balloons were released in the Southern Hemisphere to obtain information on winds over remote regions of the Southern Ocean, where conventional balloon data were unavailable. The balloon trajectories provided valuable information on the circulation of the upper troposphere but their impact on weather prediction has not been so marked. As mentioned earlier, a major difficulty with a Lagrangian system is that the observer has no control over the position of the balloon and so the velocity information may not be in the area required by the meteorologist. Another problem was that the formation of ice on the balloon's surface resulted in a high failure rate during the research programme.

A relatively recent technique for obtaining synoptic scale wind information over remote areas is by measuring the velocity of clouds. High-resolution cloud pictures, in the visible and the infra-red, from geostationary satellites allow the regular monitoring of patches of cloud, on an hourly basis, from which a time-averaged wind speed and direction can be obtained. Furthermore, because many cloud elements can be monitored, a pattern of the large-scale wind circulation can be obtained from a few photographs. The main drawback of the technique is that it is not easy to determine the exact height of the clouds. To obtain the height of the clouds, additional information on the vertical temperature profile and the cloud-top temperature is required.

In the above discussion, it is clear that the main requirement for wind observations has come from the desire to obtain global wind data, in order to understand and predict the weather. The ascents of pilot and radiosonde balloons, coupled with satellite techniques, now provide regular daily observations of wind at all heights of the troposphere and the stratosphere.

In the ocean, it is only in the last decade that measuring systems have become reliable and frequent enough to obtain time-dependent maps of, admittedly, small parts of the ocean. As noted previously, only recently have current meters been widely deployed in the ocean and prior to this, most direct information on the large-scale circulation was obtained by Lagrangian measurements as opposed to Eulerian measurements. In the late eighteenth and the nineteenth centuries, with the development of colonial trade routes, information on surface currents was invaluable for minimizing sailing times. Major Rennell, in 1778, was the first to determine the surface currents of the Atlantic Ocean. Providing that the position of the ship, its heading and speed are accurately known, the surface current can be calculated. Later, Maury (1853), the superintendent of the Naval Observatory in Washington, used this technique to produce the first global charts of surface currents. These charts are still the most commonly presented data on surface currents and, with the aid of improved position-fixing systems for shipping, using satellite navigation aids, the technique is still valuable for determining large-scale flow patterns, especially where the currents are strong. Another earlier Lagrangian technique, popular for the determination of currents in the nineteenth and early twentieth century, was the drift bottle. Its main disadvantage was that it gave no information about the trajectory between the launch and the pick-up point and a current determination would always be the minimum velocity of the flow between the two points. Today, drift cards and bottom drifters are still used to determine flow patterns over small distances, particularly in estuarine environments.

The major revolution in modern oceanography has come with the development of remotely tracked ocean buoys. The concept of an ocean buoy system is quite simple but it is only recently, with improved electronic systems, telemetry and satellite navigation, that they have come into widespread use. A surface drifting buoy consists of a transmitter with batteries and an antenna for radio location by satellite, aircraft or by a shore station. It is desirable to have most of the volume of the buoy below the water to reduce the direct effect of wind stress. Second, they usually have a drogue, or parachute, at a depth of 20–30 m to enable the drifter to move with the ocean current rather than with the direct wind-induced current. Many surface drifters also record the surface atmospheric pressure and temperature. Some have a thermistor chain attached which enables the sub-surface temperature profile to be monitored (Figure 6.1(*d*)). Surface drifters are less expensive than current moorings and they also give more information on the large-scale flow patterns than an array of current meters. They can be tracked relatively easily for many months, and sometimes for years, and are ultimately limited by their batteries' duration. In 1979 a large number of surface drifters were placed in the Southern Ocean in order to monitor the surface meteorology in this data-sparse region during the First Global GARP Experiment (FGGE). In addition these drifters provided invaluable data on the surface currents over a 2 year period.

It was recognized by Stommel and other oceanographers in the 1950s that a drifting buoy could also yield important information on deep-ocean flows, which were believed to be difficult to measure by conventional techniques. John Swallow, working at the National Institute of Oceanography, U.K., was the first to design a deep-ocean float which could be tracked by a ship using hydrophones. His success lay in the observation that a body which is less compressible than sea water will gain buoyancy as it sinks. Furthermore, if the excess weight of the body at the surface is small, then at a given depth it will become neutrally buoyant and no further sinking will occur, and thus this float would become the oceanographic equivalent of the constant-level balloon in the atmosphere. The float consists of a 6 m length of aluminium tube, half of which is used for buoyancy and the other half for housing an acoustic transmitter and batteries. The original float weighed about 10 kg and weights were added to obtain a negative buoyancy of 38 g at the surface, thus allowing the float to become neutrally buoyant at a depth of about 1000 m. Two hydrophones, one at each extremity of the research ship, were used to determine the distance and direction of the float. One of the difficulties is that it is not possible for a ship to track more than a few floats at any one time. Furthermore, the ship is totally involved in locating the floats and this is a very expensive operation. A subsequent development, by Rossby *et al.* in the U.S.A. in the early 1970s, involved the location of neutrally buoyant floats in the sound channel, between 1000 and 1500 m, which enabled continuous tracking of large numbers of floats by remote, shore-based hydrophones up to 1000 km away. These SOFAR floats have been located in the western Atlantic for over a decade and a considerable quantity of information on large-scale ocean circulation has now been collected, as evidenced by figure 6.3. It is important to realize that even neutrally buoyant floats only give a picture of the horizontal Lagrangian flow at a specified depth and that they can give no information on the vertical velocity field. Second, a large density of floats is required to give a reliable, statistical picture of the flow field. For example, in the MODE 1 (Mid-Ocean Dynamics Experiment), 20 floats were required to adequately monitor a 300 km diameter region of the ocean west of Bermuda.

Recently, specialized floats have been designed to measure vertical as well as horizontal Lagrangian velocities. This has been achieved by placing fins on the sides of the float so that it rotates in proportion to the vertical flow. Such floats have been used successfully for determining vertical motion in both the Caribbean and Mediterranean Seas.

In the meteorological context, both radiosondes and dropsondes (i.e. radiosondes parachuted from aircraft) give a measurement of the vertical variation of horizontal velocity, which is useful for the detection of large local velocities associated with jet streams and internal waves. When used in conjunction with other wind measurements from other radiosonde stations, a three-dimensional picture of the horizontal flow field can be

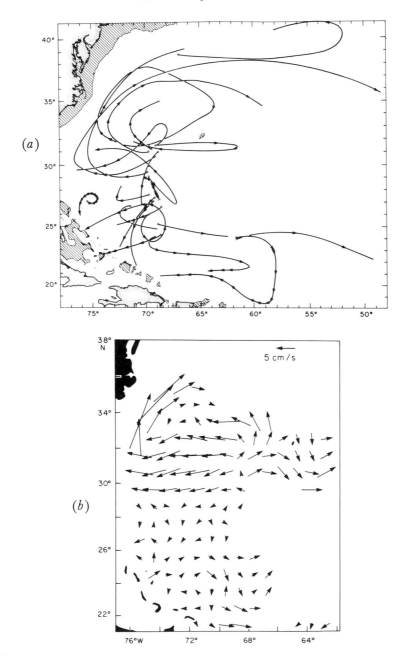

Figure 6.3. (a) A spaghetti plot of 700 m SOFAR float trajectories. Arrows are 100 days apart. Note the high velocity of floats caught in the Gulf Stream. (b) The mean equivalent Eulerian circulation estimated from 700 m SOFAR float data. (Reproduced from H. T. Rossby et al., p. 68 and 70, figures 1 and 3 in Eddies in Marine Science *edited by A. R. Robinson, with permission of Springer-Verlag.)*

obtained. An oceanographic equivalent of the dropsonde is the velocity profiler, which is simply an acoustic float with temperature and pressure sensors. It is allowed to sink slowly over an array of hydrophones on the bottom. As with the neutrally buoyant floats, the time delays between the float emission and the hydrophone reception allow the float's position to be determined to an accuracy of ∓ 1 m. By differencing the position plot, the horizontal velocity can be obtained with an accuracy of ∓ 2 cm s^{-1}. Figure 6.4 shows some results obtained by Luyten and Swallow using a vertical profiler on the equator in the Indian Ocean. The considerable vertical variation of equatorial currents surprised both experimental oceanographers, using conventional current-meter moorings, and theoretical oceanographers. It demonstrated, yet again, just how much remains to be observed and understood about the ocean circulation.

All the Lagrangian techniques described above only measure a part of the three-dimensional flow pattern and therefore represent only quasi-Lagrangian methods. However, it has long been recognized that a chemical or radioactive tracer in a turbulent environment like the ocean will move with the individual parcel of water into which it is inserted, provided that it is in a sufficiently low concentration to leave the physical properties of the water mass unaltered. In Chapter 5 it was shown how naturally

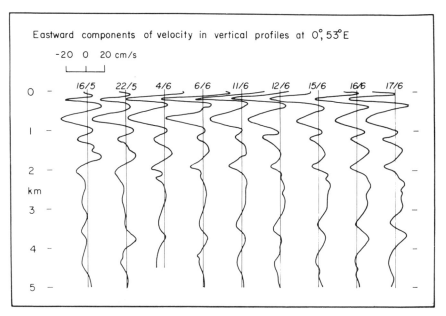

Figure 6.4. Horizontal component of velocity obtained from a velocity profiler in the equatorial Indian Ocean. (Reproduced from J. R. Luyten and J. Swallow (1976), Deep Sea Research, **23***(10), 1001, figure 1, with permission of Pergamon Press.)*

occurring chemical and radioactive tracers can be used to infer circulation patterns, but the monitoring and diagnosis of a Lagrangian tracer experiment are quite different to the techniques previously described. Consider the evolution with time, of a patch of dye on the ocean surface. The dye spreads in all directions as the result of a large number of small-scale fluctuations in the surface velocity induced, perhaps, by the surface wind or by small eddies in the flow. This is an example of diffusion. Furthermore, the centre of gravity of the dye patch will show a translation. Thus the dye gives two types of information:

(i) The area spread of the dye gives information on the mixing by time-dependent currents.

(ii) The movement of the centre of gravity of the dye patch gives information on the mean current.

To monitor a tracer experiment, it is necessary to sample both spatially and with time. Time-lapse photography of the dye patch, from a ship or an aircraft, can give useful information over a small surface area, but below the surface, photography can only be used in the clearest waters. Fluorimeters, towed along beneath the surface by a ship, can be used to measure dye concentrations accurately and quickly. A useful, non-toxic dye is rhodamine-β and it can be detected at concentrations of 1 part in 10^{10} by fluorescence measurements. The most frequent use of dye measurement is in local areas such as coastal seas and estuarine environments where it is used to study the dispersal of pollutants.

Chemical and radioactive tracers require more time-consuming onboard and shore-based techniques to measure concentrations. Care has to be taken to prevent the contamination of samples during collection and transfer to the laboratory, but such tracers do have the advantage of being able to yield high-precision results. Tritium, produced by atmospheric hydrogen-bomb tests in the late 1950s and early 1960s, has been used to trace both horizontal and vertical ocean currents over large areas of the North Atlantic. In the waters around the British Isles, circulation has been determined by the regular monitoring of caesium-145 which is a low-level radioactive waste product from the nuclear reprocessing plant on the Cumbrian coast.

6.2. Climate and seasonal circulation

Climate and seasonal time-averaged circulation of the atmosphere

Care has to be taken in the interpretation of the time-averaged pictures of circulation systems as complex and variable as the ocean and the atmosphere. For the purposes of clarity, the atmospheric mean circulation will be defined as the climatic monthly average taken over a period of 5–10 years. For surface observations, a 30 year average is usually taken as the climatic average but there is no *a priori* reason to suppose that this will be more representative of the circulation than a 10 year average. Figure 6.5 shows the longitudinally averaged winds as a function of height and latitude for the months of January and July, obtained from daily observations of the atmosphere from weather stations, ships, aircraft and balloons. The predominant feature in both hemispheres is a region of high wind speed, reaching 30 m s^{-1}, in the upper troposphere between 30 and 40° latitude.

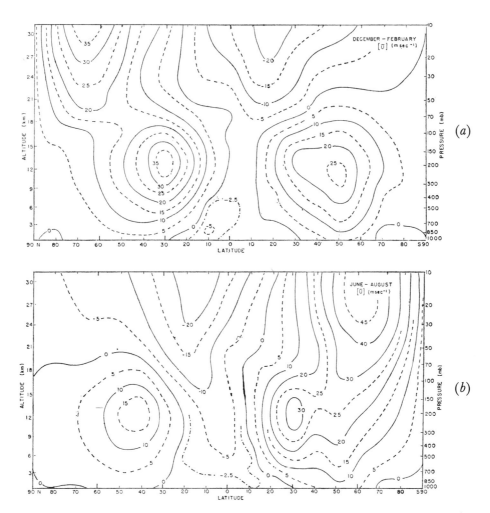

Figure 6.5. Mean zonal wind for (a) December–February, and (b) June–August. Units: m s^{-1}. Positive values denote eastward flow. (Reproduced from R.E. Newell et al., p. 65, figures 14a and b, in The Global Circulation of the Atmosphere *edited by G.A. Corby, with permission of the Royal Meteorological Society.)*

These climatological, eastward-flowing jet streams are the most energetic part of the tropospheric circulation. It is noted that the Northern Hemisphere sub-tropical jet stream moves poleward in the summer and weakens by about 50%. A similar poleward movement occurs in the Southern Hemisphere, but the seasonal variation in intensity is smaller. The magnitude of the jet stream is related to the horizontal meridional temperature gradient and this is discussed in Chapter 7. The mean jet flow is dominated by eastward winds throughout most of the upper troposphere

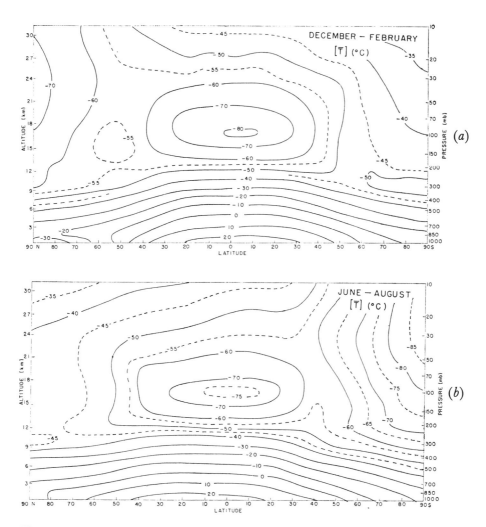

Figure 6.6. Mean temperature for (a) December–February, and (b) June–August. Units: °C. (Reproduced from R. E. Newell et al., p. 65, figures 9a and b, in The Global Circulation of the Atmosphere, *edited by G. A. Corby, with permission of the Royal Meteorological Society.)*

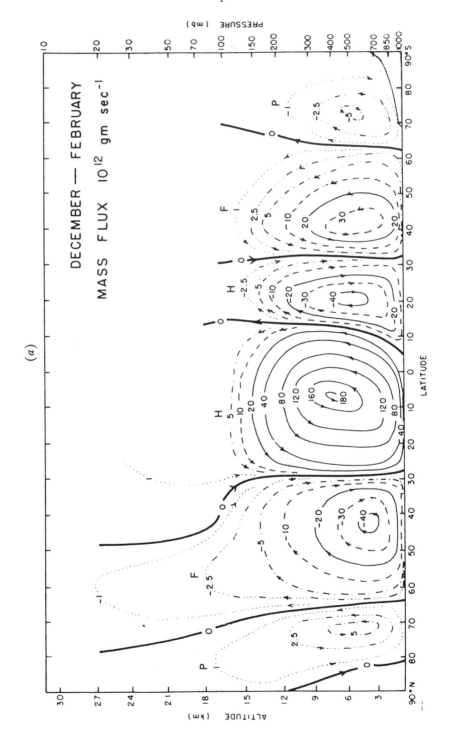

(a)

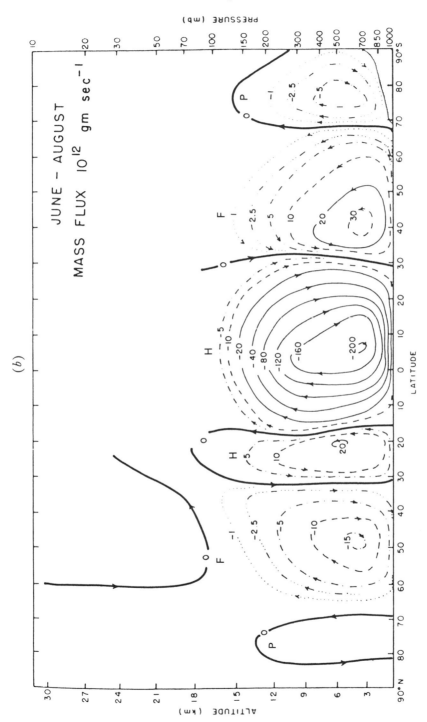

Figure 6.7. *Mass flux in the meridional circulation showing Hadley cells (H), Ferrel cells (F) and Polar cells (P). (a) December–February and (b) June–August. (Reproduced from R.E. Newell et al., The General Circulation of the Tropical Atmosphere, figure 3.19, page 45, with permission of MIT Press.)*

but westward winds do occur in the tropical troposphere and near the surface in polar regions. It can be appreciated that it is necessary to have both eastward and westward winds at the surface to avoid a net wind torque on the Earth which would eventually result in a change in the Earth's rotation.

In the stratosphere there is a very important seasonal variation in winds. There is a general tendency for the eastward jets to weaken immediately above the troposphere but, in winter a strong jet develops at high latitudes with magnitudes of up to 70 m s^{-1}. This is known as the polar night jet and it is caused by the large meridional temperature gradient between the polar and the sunlit stratosphere (Figure 6.6). As discussed in Chapter 3, the temperature of the stratosphere is controlled by the absorption of solar radiation by ozone. During the polar night, in the absence of solar heating, there is a rapid decline in the temperature of the polar stratosphere. In the polar summer, the meridional temperature gradient is reversed because maximum radiation occurs at the poles, and the stratospheric wind circulation in the summer hemisphere is everywhere westwards. This situation is depicted in figure 6.5.

A different picture of the mean circulation is obtained from figure 6.7 which shows the vertical and meridional components of the flow. To help visualize this flow, the stream function is shown. The flow is along the streamlines and its magnitude is proportional to the gradient of the stream function. A large cell, known as the Hadley cell, is seen to cover the tropical region and two much weaker cells, known as the Ferrel cell and polar cell, respectively, are also visible. In the Hadley cell, air is lifted up over the equatorial zone, mostly in the ITCZ, where large amounts of latent heat are released by heavy precipitation. This air moves polewards to the sub-tropics, where it descends. During its descent the air warms by adiabatic compression and its relative humidity decreases. The circuit is completed by a return flow in the lowest 2 km of the atmosphere towards the convergence zone. This return flow is very steady and it is associated with the north-east and south-east trade-wind systems. Some of the descending air in the sub-tropics travels polewards and ascends in the middle latitudes. This ascent occurs in the region of the sub-polar low-pressure systems. A weak return flow to low latitudes completes the circuit. Also apparent in figure 6.7 is the seasonal variation in the Hadley cell associated with the tropical monsoons. One feature of the mean meridional circulation is that it is much weaker than the corresponding zonal circulation, having mean velocities of less than 3 m s^{-1}, and the maximum velocity occurs in the low-level trade winds. The vertical velocities are generally less than 1 cm s^{-1}. The mean meridional circulation is, therefore, a much weaker feature than the zonal circulation, which is related to the dominance of the equatorial–polar temperature gradient in controlling the flow.

The horizontal climatological surface flow is shown in figure 6.8 for

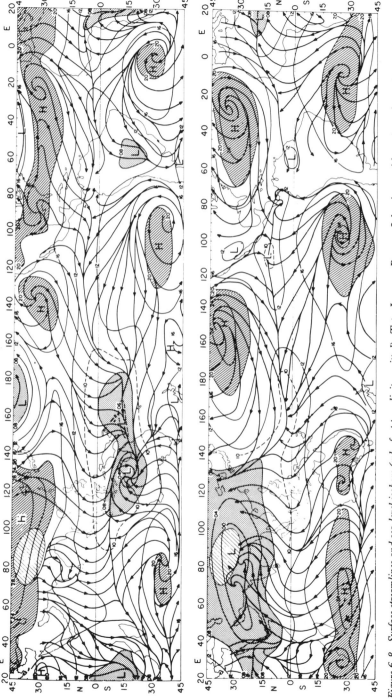

Figure 6.8. Surface streamlines and sea-level isobars (mb, first two digits omitted). Top: January. Bottom: July. Areas with pressure above 1020 mb are shaded; below 1008 mb, hatched. (Reproduced from Climate and Weather in the Tropics, *by H. Riehl, p. 14, figure 1.8, with permission of Academic Press.)*

January and July. Here the flow is depicted by streamlines, which allow a continuous flow picture to be obtained as well as showing the origin of the air masses. It can be seen that the sub-tropical high pressures are the 'sources' of the circulation. These are regions where descending air from the upper branch of the Hadley cell enters the lower troposphere and they are largely cloud-free, with only low-level broken cloud usually observed. The trade winds can be seen flowing towards the equatorial zone where they meet, from both hemispheres, at the ITCZ. These convergence zones are generally located in the Northern Hemisphere in the Pacific and Atlantic Oceans along the thermal equator (i.e. the region of maximum sea surface temperature). In the Indian Ocean and the western Pacific Ocean seasonal monsoonal variations in the wind circulation dominate the flow. The large mass flow from the southern Indian Ocean high to the Asian continent is evident in summer, from about June to September. The curving of the winds from a south-east flow in the Southern Hemisphere to a south-west flow in the Northern Hemisphere is related to the rotation of the Earth, as discussed in Chapter 7. In winter the low-pressure centre is located over the Indonesian maritime continent and covergence of air from both the Indian and Pacific Oceans can be seen. The regions of convergence indicate areas of upward motion and heavy precipitation, as already shown in figure 5.6. Weaker monsoonal circulations occur over Africa and South America.

Climate and seasonal circulation of the ocean

The major data source for the climatic determination of surface ocean currents has been ships' drift measurements. A chart, such as the one shown in figure 6.9, probably represents well over 100 years of data, obtained from merchant vessels. The main features of the chart include:

  (i) The locations of the most intense currents in the equatorial zones and along the east coasts of Japan (the Kuroshio current), North America (the Florida current and the Gulf Stream) and South Africa (Agulhas current), where peak velocities reach over $1 \mathrm{~m\,s}^{-1}$.

 (ii) The tendency of mid-ocean currents to be weaker. Surface velocities of $10-30 \mathrm{~cm\,s}^{-1}$ are recorded.

(iii) The east–west nature of the currents in the equatorial Pacific and Atlantic.

(iv) The domination of an eastward current around the globe in the Southern Ocean.

 (v) The alignment of the current boundaries, or frontal regions, in an east–west orientation.

(vi) The presence of large anticyclonic flows at approximately $30°$ latitude in all ocean basins except for the northern Indian Ocean.

(vii) The response of the surface currents in the Indian Ocean to the monsoons.

(viii) The small cyclonic circulations occurring in the north-western Atlantic, in the North Pacific near the Aleutian Islands and in the Weddell Sea.

It is unlikely that more information on surface currents will drastically alter the above description of the major features. However, below the surface it is not possible to obtain a climatic picture from direct measurements, except over small regions. It is only since the 1950s that current meters have been deployed in significant numbers and only in the past 15 years have long-lived moorings been in service. The present picture of the mean circulation at depth is, therefore, incomplete. However, hydrographic data on temperature, salinity and pressure can be used to obtain an indirect estimate of the mean current. The details of this method will be discussed in Chapter 7 but, in its simplest form, it involves the calculation of the horizontal pressure field from the density of sea water and the use of the hydrostatic equation. Except on the equator, the horizontal flow tends to follow lines of constant pressure and hence the circulation can be calculated. This flow is known as a geostrophic current. However, only relative horizontal motion can be obtained because it is not possible to calculate the absolute horizontal pressure field. The calculated relative pressure field, between the surface and 15,000 kPa (or approximately 1500 m) is shown in figure 6.10. The relative geostrophic flow can be determined from the pressure field gradient. The main features of the calculated flow are the large anticyclonic gyre circulations in the sub-tropical latitudes, which are seen clearly in the surface current pattern (figure 6.9). The intense pressure gradients in the western parts of the North Atlantic and North Pacific Oceans are associated with the Gulf Stream and Kuroshio current. It is noted that the sub-tropical gyres are located adjacent to the western boundaries of the oceans. Polar cyclonic gyres are in evidence in both the North Atlantic and North Pacific Oceans and in the South Atlantic Ocean adjacent to South America. Again, the centres of these gyres are located close to the western boundaries. At higher latitudes of the Southern Hemisphere the broad eastward flow of the Antarctic Circumpolar Current is seen. This current moves equatorwards in the South Atlantic and western Indian Ocean and then plunges polewards in the South Pacific. Close to Antarctica a double low-pressure gyre can be seen in the Atlantic sector of the Weddell Sea. It will be noticed that this indirect technique implies occasional flows through the ocean boundaries. These are regions where the geostrophic theory breaks down and a non-geostrophic boundary flow will develop to conserve mass. These boundary flows are often intense currents, like the Agulhas current off South Africa and the east Australian current.

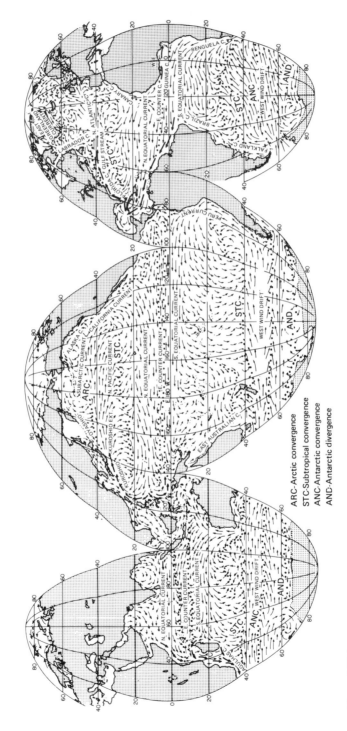

ARC-Arctic convergence
STC-Subtropical convergence
ANC-Antarctic convergence
AND-Antarctic divergence

Figure 6.9(a)

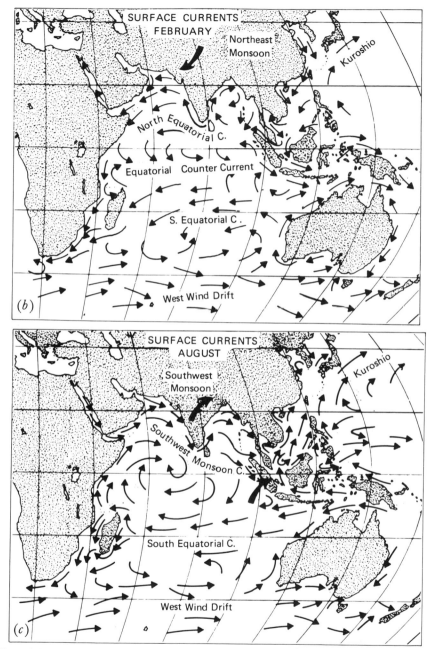

Figure 6.9. (a) Ocean-surface currents in February–March. The thick arrows indicate especially strong currents. (Reproduced from The Oceans: Their Physics, Chemistry and General Biology, by H. U. Sverdrup et al., with permission of Prentice-Hall.) (b) Monsoonal surface currents: February. (c) Monsoonal surface currents: August. (Reproduced from Oceanography: A View of the Earth, by M. Grant Gross, p. 180, figure 7.5, with permission of Prentice-Hall.)

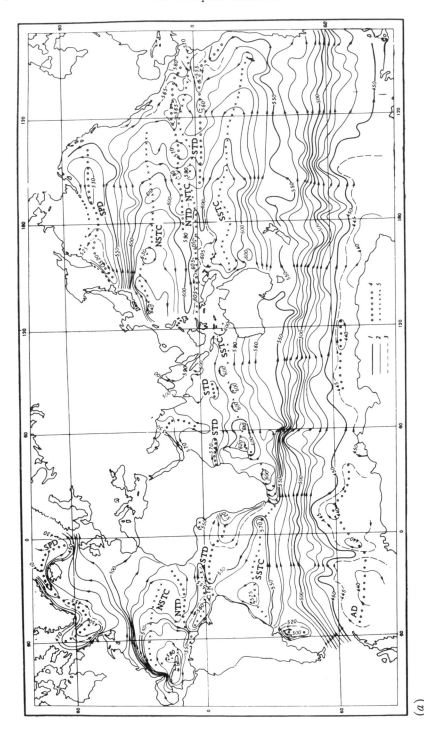

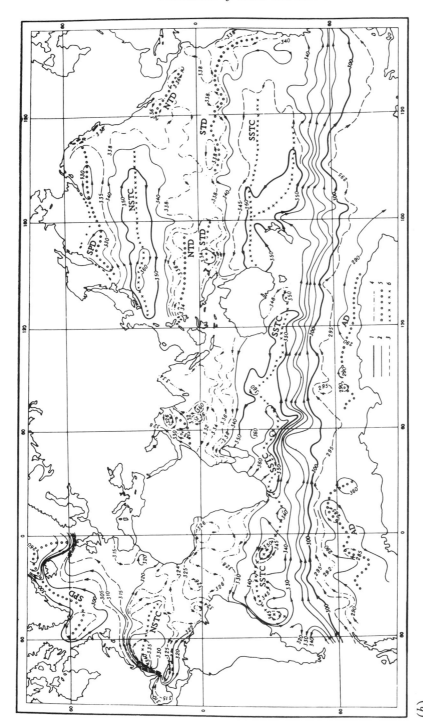

*Figure 6.10. Dynamical height (dynamic cm) (a) at the sea surface, and (b) at 500 db, with respect to the 1500 db surface. Contour interval = 10 dynamic cm. Major frontal regions and water mass boundaries: SPD = subpolar divergence; NSTC = northern subtropical convergence; NTD = northern tropical divergence; NTC = northern tropical convergence; STD = southern tropical divergence; SSTC = southern subtropical convergence; AD = Antarctic divergence. (Reproduced from V.A. Burkov et al. (1973), Oceanology, **13**, 325–332, with permission of the American Geophysical Union).*

(b)

Another important feature of the mean flow can be seen in the North Atlantic where a broad poleward flow, called the North Atlantic drift, enters the Norwegian Sea and the Arctic Basin. This returns from the Arctic Basin into the North Atlantic via the intense east Greenland boundary current. However, not all of the poleward flow is compensated by return flows between the surface and 1500 m and this implies that the sinking of water must occur for conservation of mass. This sinking occurs in parts of the Norwegian Sea, the Greenland Sea and the Labrador Sea. However, because of the bottom topography and the episodic nature of this sinking, the abyssal flow is not a continuous, steady, southward current. It tends to be most intense close to the western edge of the western Atlantic basin.

The observations of vertical motions in the ocean are, at best, scanty. Figure 6.11 shows a schematic vertical section across the Atlantic Ocean for the uppermost 800 m of the ocean, obtained from a variety of indirect methods. In the surface layers the eastward and westward wind-driven currents are evident. The surface currents produce regions of convergence at approximately $30°$ latitude in both hemispheres in the centres of the sub-tropical gyres and this results in downward motion in the upper $200-400$ m which, in turn, is partially responsible for the dip in the thermocline at these latitudes. A large fraction of this water is subsequently transported upwards and equatorwards along the main thermocline to the equatorial upwelling region, where the thermocline is often less than 100 m below the surface. In the Northern Hemisphere two small, closed cells are found between the equator and $10°N$, where downwelling and a dip in the thermocline are produced. The dipping of the thermocline induces an eastward geostrophic flow between the generally westward equatorial current and is known as the equatorial counter current.

The vertical flows in the middle and tropical latitudes of the open ocean are induced by the surface currents. In polar latitudes the sinking of water and movement along isopycnal surfaces induce vertical circulations. The equatorial upwelling zone may produce vertical velocities of up to 10^{-4} m s^{-1}, whilst sub-tropical downwelling regions have velocities of less than 10^{-6} m s^{-1}. In the deep ocean vertical motions are associated with velocities of approximately 10^{-7} m s^{-1}, although local values of 10^{-3} m s^{-1} have been recorded in very small sinking regions less than 10 km in diameter. The very small, mean vertical flows in the deep ocean imply that direct current measurements will have to be made for a very long period of time if the net flow is to be determined unambiguously. In these regions the use of tracers has proved to be invaluable in deducing the vertical flow. Figure 5.16 shows the tritium distribution in the North Atlantic about 10 years after its surface input ceased. This figure clearly shows the weakness of the vertical circulation in the Atlantic Ocean. Deep sinking is evident at high latitudes in the North Atlantic

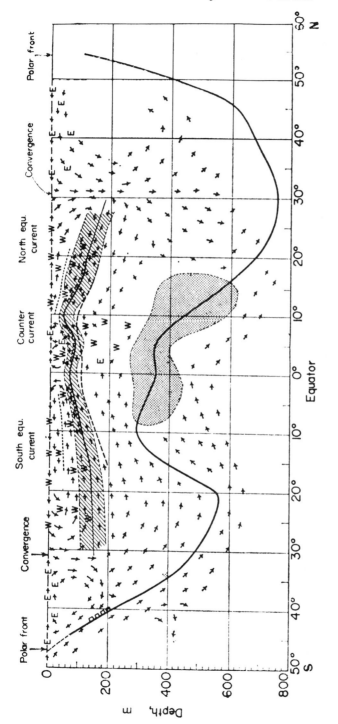

Figure 6.11. Schematic picture of structure and circulation in the troposphere of the Atlantic Ocean in the meridional direction. (Reproduced from Physical Oceanography, Vol. 1, by A. Defant, p. 600, figure 274, with permission of Pergamon Press.)

          Boundary between the warm upper ocean and the cold polar ocean.

          Tropical–subtropical thermocline

          Layers of extremely low oxygen contents ($< 1·5\ cm^3\ 1^{-1}$).

          Position of tropical–subtropical salinity maxima.

W, E   *Zonal velocity component (W toward west, E toward east)*

but, in general terms, all of the remaining tritium is confined to the
waters above the main thermocline. The upward bowing of the distribution is evident in the equatorial zone where upwelling occurs.

From this brief resumé of the mean ocean circulation it is clear that
more remains to be discovered than is known at present. However, by
using different indirect techniques of determining current flow, it has
been shown that the climatological flow is much weaker in the deep ocean
than at the surface. The surface flows are determined, to a substantial
degree, by the surface wind distribution and seasonal changes of the wind
circulation can result in changes in the surface circulation, as seen in the
Indian Ocean. The deep flows are not directly determined by the surface
winds and are more dependent on the vertical sinking of polar waters (the
thermohaline circulation), the horizontal density gradient and the
geometry of the ocean basin.

6.3. Scales of motion in the atmosphere and ocean

Figure 6.12 shows four individual time series of wind speed measured by
an anemometer. The first series shows the fluctuations in velocity
covering a period of 1 min, the second shows 1 min average values for a
period of 1 hour, the third shows 1 hour average values for a period of
4 days and the fourth shows 1 day average values for a period of 1 year.
The main conclusion that can be drawn from these four time series is that
fluctuations in wind velocity occur on a whole range of time scales, from
a few seconds to years. Time series, not dissimilar to those of the atmosphere, can also be obtained for the ocean. A meteorologist, or an oceanographer, has to be aware of these fluctuations and their magnitudes. For
example, a synoptic meteorologist who wanted to obtain wind
information to discern large-scale weather systems would be interested in
obtaining 10 min or hourly averaged winds, in order to remove the high-
frequency variations which have little direct relevance to large-scale flow
patterns. On the other hand, a meteorologist, who wanted to know how
the smoke from a chimney was going to behave in the immediate vicinity
of the discharge, would be interested in the high-frequency fluctuations
in the wind, because it would be the gusts of wind which would be
responsible for the mixing of the smoke and the spread of the plume. A
climatologist would probably average winds over a month, a season or
even a few years in order to ascertian long perid fluctuations and trends.

A useful way to represent these fluctuations is to convert the time
series shown in figure 6.12 into time frequency spectra, as illustrated in
figure 6.13. The simplest approach to this kind of analysis is to consider
a time series as the summation of a series of sine and cosine functions having a range of discrete frequencies. The highest frequency is $\pi/\Delta t$, where

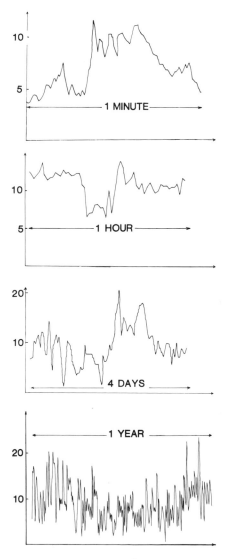

Figure 6.12.   Four time series of wind speed, $(m\,s^{-1})$: 1 minute, 1 hour, 4 days and 1 year.

Δt is the sampling period of data, and the lowest frequency is $2\pi/T$, where T is the period of the observations. Thus a time series can be represented by:

$$u(t) = u_0 + \sum_{n=1}^{N} a_n \cos nt + b_n \sin nt \qquad (6.5)$$

where u_0 is the mean value of the time series. The coefficients a_n and b_n

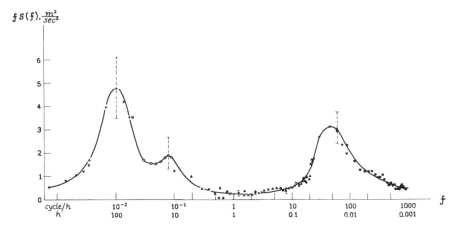

Figure 6.13. Frequency spectrum of the horizontal wind velocity. Some experimental points are shown on the graph. (Reproduced from J. van der Hoven (1957), J. Met., 14, 161, figure 1, with permission of the American Meteorological Society.)

can be determined from the following relationships:

$$a_n = \frac{2}{T}\int_0^T u(t) \cos\left(\frac{2\pi nt}{T}\right) dt$$

$$\tag{6.6}$$

$$b_n = \frac{2}{T}\int_0^T u(t) \sin\left(\frac{2\pi nt}{T}\right) dt$$

where T is the length of record.

A Fourier decomposition reveals two categories of information. First the modulus of the coefficients, given by $\sqrt{(a_n^2 + b_n^2)}$, yields a measure of the amplitude of the fluctuations at frequency n, whilst the ratio of the coefficients, $\tan^{-1}(a_n/b_n)$, determines the phase in radians, of the fluctuations at frequency N. Though the Fourier method, and also other spectral methods using finite data, produce a discrete set of frequencies, there is a tendency in the literature to show the spectra as continuous, as in figure 6.13. As can be seen from this figure, there is no single frequency which dominates the spectrum, although there are certain frequencies where the fluctuations are large. In this atmospheric spectrum there is a peak of fluctuations with a period of 1 min associated with gusts produced by convection processes. There is also a tendency for high energy fluctuations at 1 day which is caused by diurnal variations in the wind, whilst between the period of 1 min and 1 day the fluctuation energy is smaller. For periods longer than a day there is a dramatic increase and a maximum occurs at a period of a few days. This is due to the large-scale synop-

tic variability of the atmosphere. Transforming the time series into a
frequency spectrum thus makes it possible to perceive different time scales
of motion in the atmosphere.

Figure 6.14 shows a frequency spectrum obtained from the analysis
of readings from a moored current meter at a depth of 40 m on the continental shelf. It is evidently quite different from the atmospheric
spectrum, having three identifiable peaks. Two peaks arise form the semi-
diurnal tide (period 12·4 hours) and the diurnal tide (period 24 hours)
and the third peak is an inertial oscillation having a period of 18·5 hours.
The inertial oscillation will be discussed in Chapter 7, but it is essentially
a function of latitude and the rotation of the Earth. Its period, T, is given

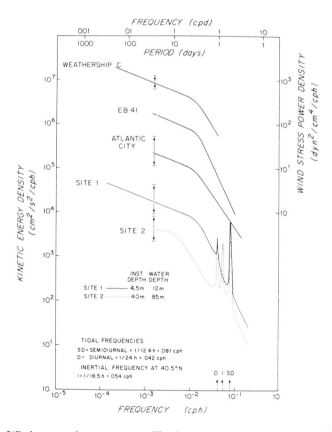

*Figure 6.14. Wind stress and current spectra. The three upper curves represent the wind stress spectra
for ocean weathership C, environmental buoy EB41, and Atlantic City, New Jersey. The two lower
curves are the kinetic energy spectra obtained at a nearshore site off New Jersey and a deeper site located
near the shelf break south of New England. Note the tidal and inertial peaks in the spectrum.
(Reproduced from R. A. Beardsley, chapter 7, p. 217, figure 7.8 in* Evolution of Physical
Oceanography, *edited by B. A. Warren and C. Wunsch, with permission of MIT Press.)*

by:

$$T = \frac{\pi}{\Omega \sin \phi} \tag{6.7}$$

where Ω is the rotation rate of the Earth and ϕ is the latitude. At longer periods the spectrum shows no obvius peak but does display a large amount of variability. This suggests that a variety of mechanisms may be causing low-frequency time variability in the ocean. In this particular example both wind and current spectra have a similar shape and this suggests that these long-period ocean currents may be directly generated by the wind.

Of course, the time variability is only one of several possible dimensions in which to analyse motions. The spatial equivalent is the wavenumber spectrum, the wavenumber being the inverse of wavelength. In this case, unlike the infinite time series, there are quite definite limits to the spectrum which are defined by the finite size of the ocean basins and the Earth's atmosphere. Figure 6.15 demonstrates the wavenumber frequency spectrum of the wind, obtained from the analysis of the large-scale atmospheric flow. It can be seen that a large part of the variability is on very long scales, with peak energy in waves which have 5–8 cycles around

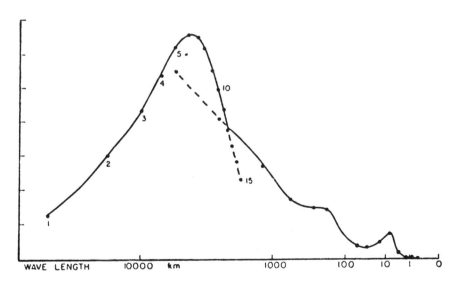

Figure 6.15. Wavelength spectrum of wind fluctuations in temperate latitudes. Numbers indicate circumpolar wavenumbers. Dashed segments are extensions of results of separate studies beyond point of intersection. Area under segment of curve is proportional to eddy kinetic energy in corresponding band of the spectrum. (Reproduced from E. N. Lorenz, p. 11, figure 4 in The Global Circulation of the Atmosphere, *edited by G. A. Corby, with permission of the Royal Meteorological Society.)*

the Northern Hemisphere. The lowest wavenumbers (1–4) are long Rossby waves and they are associated with the distribution of continent and ocean and with topography. Higher wavenumbers, particularly between 8 and 10, are associated with synoptic weather systems, which have space scales of a few thousand kilometres. At higher wavenumbers (i.e. shorter wavelengths) the energy of the spectrum drops very quickly, indicating that the large-scale systems are, energetically, the dominant flow in the atmosphere. As will be shown later, this is not true of the ocean basins where it is found that the energetically dominant space scales are associated with mesoscale circulations having wavelengths of approximately 100 km.

6.4. Time-dependent motion

Time-dependent atmospheric motion

It has been shown that time-dependent motions occur on a variety of space and time scales. The largest scales of motion constitute the most energetic and the longest time-scale phenomena in the atmosphere. Figure 6.16 shows the trajectory of a constant-level balloon at 150 mb (or approximately 15 km) observed over a period of 23 days. If the flow was purely zonal and constant in time, concentric circles would be anticipated with the balloon repeating the trajectory on each circuit, but considerable variations in time and space are observed. On the last circuit of the hemisphere there are two large excursions over South America and the eastern Atlantic Ocean which were not apparent on the previous circuit. These deviations from circular motion are caused by Rossby waves. The number of these Rossby loops in the circumpolar flow can vary from one or two to as many as six or seven waves. If they were sinusoidual waves, a spectrum such as the one shown in figure 6.15 would indicate their wavenumber. For example, a wavenumber 1 flow would be a single wave around the hemisphere whilst a wavenumber 4 flow would indicate four wave loops round the hemisphere. The central axis of the waves is often associated with jet streams and, unlike the climatological sub-tropical jet stream with mean winds of 30 m s^{-1}, wind speeds of up to 100 m s^{-1} can occur. The observation and forecasting of these jet streams are important priorities for aircraft operators. The low wavenumber patterns tend to move only slowly with respect to the Earth's surface, with the circumpolar flow moving through the pattern. Occasionally the Rossby wave pattern may become stationary for a few days, or even weeks, and then it is known as a 'blocked flow'. The waves are generated by a dynamic instability of the strong circumpolar flow, which grows on a time scale of a few days. However, the waves may be enhanced by large mountain

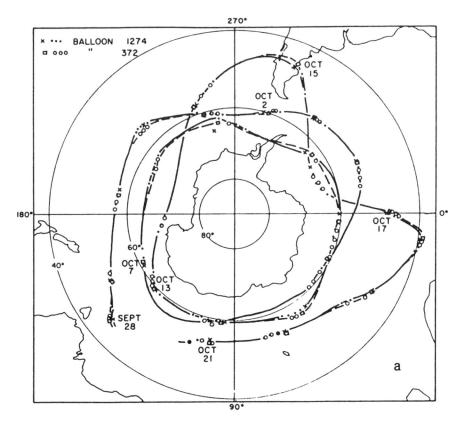

Figure 6.16. A portion of the trajectories of TWERL balloons 1274 and 0372. Successive positions on the dates indicated are also shown. (Reproduced from The Twerle Team (1977), Bull. Am. Met. Soc., 58, 942, figure 6a, with permission of the American Meteorological Society.)

barriers such as the Andes or the Rocky Mountains and by the thermal contrast between ocean and continent.

Embedded within these large Rossby wave loops are the familiar, synoptic scales cyclones and anticyclones which have length scales of the order of 1000–3000 km and time scales of 1–5 days. There are typically 20–25 cyclones and anticyclones in the lower part of the atmosphere (i.e. below 5 km) in the Northern Hemisphere at any one time. There tend to be larger numbers of cyclones than anticyclones but the latter tend to have larger horizontal scales. The extra-tropical, or middle latitude, cyclone is also associated with stronger wind circulation than the anti-cyclone. The positions of the cyclones and anticyclones do not tend to be random with respect to the upper air flow. The cyclones form on the eastern side of the trough of the Rossby wave, whilst anticyclones are generally located within, or slightly downstream of, the ridge. Within a stationary or 'blocked' flow, the anticyclones and cyclones will grow and decay repeatedly over the same regions thus resulting in persistent regions of 'good' and 'bad' weather. The surface pressures of anti-cyclones are generally in the range of 1020–1050 mb, whilst extra-tropical cyclones have a range of 930–1010 mb. Both highest and lowest pressures tend to occur in the winter hemisphere. Vertical motions within synoptic systems tend to be between 1 and 10 cm s^{-1}.

Within the extra-tropical cyclones are frontal systems which range from 50 to 200 km in width. These frontal regions tend to spiral out from the cyclone's centre and they may extend for many thousands of kilometres and even into another cyclonic system. Frontal regions are boundaries separating air masses and are, therefore, associated with large temperature gradients, especially in winter. Temperature changes are usually between 5 and 10°C, but occasionally, over continental regions, temperature changes of up to 20°C are observed. These frontal regions are often dynamically active and they are associated with upward motion and continuous precipitation, as shown in figure 6.17. The strongest up-ward motion in an extra-tropical cyclone is usually associated with the frontal systems and it may reach values of 0·5 m s^{-1} in intense fronts.

The next important scale in atmospheric motions is the mesoscale which includes a variety of phenomena ranging from shower clouds, thunderstorms and sea-breeze circulations to larger systems such as squall lines and rain bands. The former group of phenomena has scales of 1–10 km and time scales of the order of 1 hour whilst the latter may reach 100 km in size and persist for up to 1 day. The squall line is often associated with a line of thunderstorms and strong, gusty winds and it is common in tropical latitudes. Rain bands are regions of heavy precipita-tion which exist in, and propagate through, larger mid-latitude frontal systems. The tropical cyclone, typhoon or hurricane is a large mesoscale system, of the order of 100 km in diameter, associated with intense winds of up to 50 m s^{-1} and heavy precipitation. Vertical velocities in mesoscale systems may vary from 1 m s^{-1} in a tropical cyclone or a sea-breeze cir-culation to more than 10 m s^{-1} in intense thunderstorms.

Smaller-scale systems, less than 1 km in size, tend to last no longer than a few minutes. The most ferocious small-scale phenomenon is the tornado which, although less than $\frac{1}{2}$ km in diameter, may attain circula-tion velocities of between 50 and 100 m s^{-1}. Vertical velocities tend to be of a similar order of magnitude. Waterspouts are small-scale funnel clouds which form over the sea. They are usually less intense but more numerous than tornadoes. Both tornadoes and waterspouts tend to have a cyclonic rotation which is derived from the circulation of the parent thunderstorm cloud. Small-scale whirlwinds, including the dust devil, may attain wind speeds of 10 m s^{-1} in a column a few tens of metres wide and about a hundred metres high; persisting for a couple of minutes.

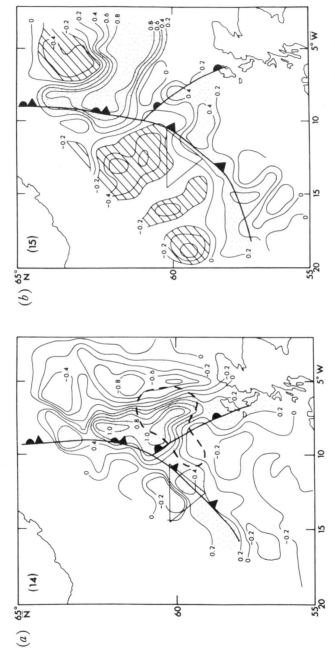

Figure 6.17. (a) Vertical velocity (W: cm s⁻¹) distribution at the top of the atmospheric boundary layer at 23.11 hours G.M.T. on 30 August 1978. Values more than $0\cdot2$ cm s⁻¹ are shown stippled; dashed line encloses a region where the rain rates exceeded $0\cdot4$ mm hour⁻¹. (b) Vertical velocity (W: cm s⁻¹) 2 hours later. Values more than $0\cdot2$ cm s⁻¹ are shown stippled, values less than $-0\cdot2$ cm s⁻¹ are hatched. (Reproduced from T. Guymer et al. (1983), Phil. Trans. Roy. Soc., A308, 266, figure 14, with permission of the Royal Society.)

They are formed in the surface layer when the prevailing wind is light and the surface is being intensely heated. The temperature gradient close to the surface becomes very large, or superadiabatic, and the surface layer becomes unstable. The strong upward acceleration results in the spectacular growth of a vortex. Other small-scale phenomena include the circulation within a cumulus cloud and the local eddying motion around buildings and obstructions. Beyond the small-scale phenomena are the microscale motions associated with the final dissipation of kinetic energy into thermal energy by viscosity.

Time-dependent oceanic motion

Ever since surface current maps were first produced in the nineteenth century the variability of ocean currents has also been recognized. Indeed, on surface charts such as figure 6.9 information is included on the steadiness of the currents, in addition to their mean speed and direction. As detailed in Section 6.3, both current meter and float measurements have indicated the presence of variability not only in the surface layers of the ocean but also in the deepest abyssal layers. A small part of the variability is accounted for by predictable frequency variations such as tides and by inertial motion. Reference to figure 6.14 will show that the majority of variations occur on time scales greater than a few days, with a peak energy between 50 and 200 days. Beyond this time scale there is an indication of a reduction in energy but suitable time series (i.e. time series sufficiently long to distinguish these low-frequency currents) are rare. However, there are seasonal fluctuations in currents such as the Gulf Stream and, more spectacularly, the Somali current, which reverses with the monsoon, and thus variations at the annual period may be anticipated. On even longer time scales low-frequency wind forcing, from interannual and climatic variations, should be expected to produce very long period fluctuations in the ocean but, as yet, such fluctuations have not been clearly distinguished.

The first recognition of the variability of the deep ocean came with the Aries experiment west of Bermuda in 1959 when 72 Swallow floats were individually tracked for 4–10 days over an experimental period of 14 months. It had been expected, from indirect measurements of hydrographic properties, that the floats would travel northwards with a steady velocity of approximately 1 cm s^{-1}. However, the floats drifted away with velocities of 10 cm s^{-1} in all directions, much to everyone's surprise. From the apparently random pattern of the buoy tracks it was possible to deduce that energetic eddies with length scales of 50–100 km were present. Subsequent to this early experiment, many Swallow and SOFAR floats have been tracked and a more general picture of the mean deep-ocean flow has emerged. Figure 6.18 shows a series of float tracks

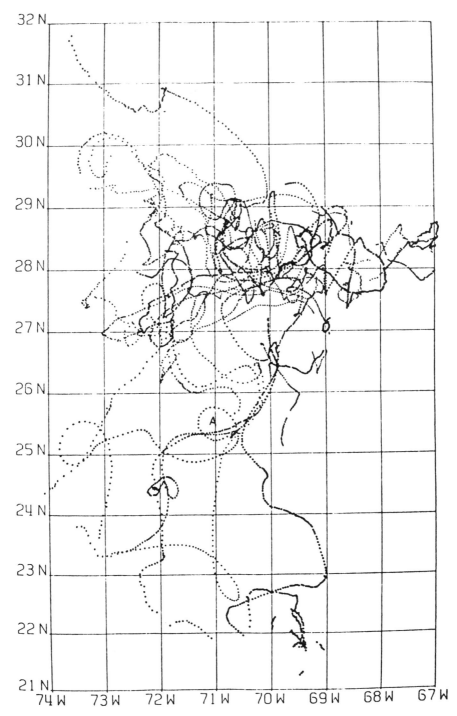

Figure 6.18(a).

obtained between September 1972 and December 1974. These plots are popularly known as spaghetti diagrams—for obvious reasons. It can be seen that floats are caught in loops or eddies for periods of a few weeks. A float resides in eddy A, which is about 60 km in diameter, for about 27 days which gives a mean speed of approximately 10 cm s^{-1} around the eddy. The floats do not show totally random behaviour because there is

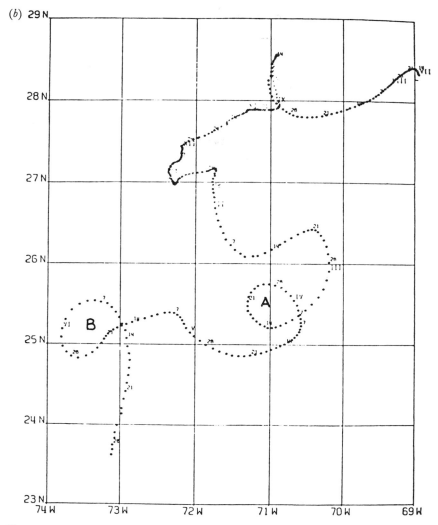

Figure 6.18. (a) Spaghetti diagram in which float trajectories between 28/9/72 and 31/12/74 are superimposed. The dots along the tracks are at 1 day intervals. Note the non-random orientation of float paths. (b) The trajectory of one float from 11/7/73 to 30/6/74. It is caught in eddy A for 27 days and in eddy B for 28 days. (Reproduced from H. T. Rossby et al. (1975), J. Marine Research, 33, 366 and 369, figures 6 and 10).

a steady drift to the west of 1 cm s^{-1} by all of the eddies. There is also a north–south flow in the region. Current meter data from moorings in the same region are shown in figure 6.19 and show current speeds of between 20 cm s^{-1} close to the surface to 10 cm s^{-1} near the bottom. The

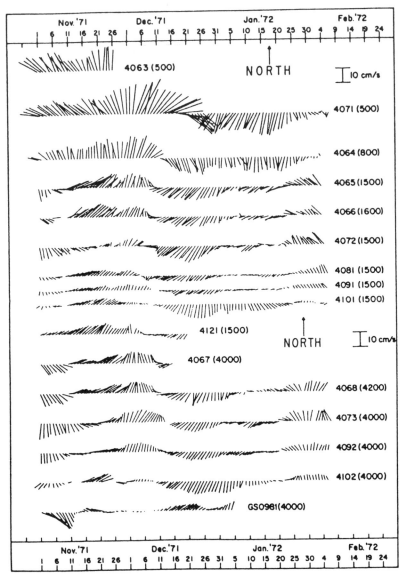

Figure 6.19. *Vector time series from November 1971 to February 1972 in MODE area in the western Atlantic. Numbers in parentheses are nominal depths in metres. (Reproduced from W. J. Gould et al. (1974),* Deep Sea Research, **21**, *916, figure 2, with permission of Pergamon Press.)*

variability of the current has periods ranging from 3 to about 6 weeks, which is typical of mesoscale eddies. These measurements, taken a decade ago in a relatively small region of the Atlantic Ocean, have shown that not only is there variability in the ocean, but also the current variability is about one order of magnitude larger than the mean velocities of the flow. This is due to mesoscale eddies and thus the kinetic energy/unit volume of the eddy is probably about 100 times larger than that of the mean circulation. Similar studies in other regions of the North Atlantic Ocean and in other ocean basins are revealing energetic, mesoscale eddies but, as yet, a detailed picture of the eddy field for all the ocean basins has not emerged. Long-term float measurements, which have been averaged over 100 days to remove the eddy field, have shown details of the large-scale circulation. Figure 6.3 clearly shows the trajectories of floats in the anticyclonic flow of the large sub-tropical gyre (i.e. clockwise, in the Northern Hemisphere). A few floats have been caught up in the intense jet-like flow of the Gulf Stream. The velocity of one such float has a mean velocity of 25 cm s^{-1} in a 100 day interval.

Other important scales of variability exist in the ocean. On the larger scale, seasonal and long-period wind variations can give rise to time variations in currents, such as those observed in the Indian Ocean. Only recently have long-period current measurements been obtained but they have already shown that remarkable oscillations in supposedly steady currents do exist. In 1974 current moorings on the equator showed significant 15 day oscillations in the flow of the equatorial undercurrent. The meanders are apparently the result of a hydrodynamic instability of the undercurrent and it has also been found that it varies on an interannual time scale, in response to changes in wind circulation. In 1982–83 the equatorial undercurrent in the eastern Pacific Ocean virtually disappeared for a period of 2 months as a result of a reversal in the surface wind from westwards to eastwards. Meanders in surface equatorial currents, about 1000 km long, have been observed by infra-red satellites measuring sea surface temperatures. Furthermore, large eddies have recently been discovered in the Somali current during the south-west monsoon. These eddies are rather larger than the mesoscale eddies described previously. They are about 500 km in diameter and tend to be slower moving.

On smaller scales, less than 100 km, surface frontal systems are observed which are energetic, for their scale. Fronts form around coastal upwelling features and form the boundary between the cold upwelled water and the warmer surrounding water. They are also found along the edge of the continental shelf break at the boundary between oceanic water and the shelf water. Convergent and divergent surface flows will also tend to result in frontal discontinuities and may be associated with surface wind patterns as, for example, the Antarctic divergence and the sub-tropical convergence zones. Such flows may also be associated with a developing eddy such as a Gulf Stream ring, where a frontal discontinuity develops about the edge of the ring, separating the warm Sargasso Sea water from colder, fresher shelf water.

Small regions of deep convection have been discovered recently in the Mediterranean and Labrador Seas and result from intense surface cooling. The convection occurs on horizontal scales of 10 km and may extend down to the ocean bottom thus providing a source of deep, cold water. Measurements with a vertical current meter have shown vertical velocities of 10^{-3} m s^{-1} in these convection regions. Such velocities are of the order of 10,000 times larger than the mean vertical velocity of the general ocean circulation, as inferred from tracers. However, these deep convective events are relatively infrequent and very localized.

The influence of the Earth's rotation on fluid motion

7.1. An introduction to the Earth's rotation

The Earth is a rapidly rotating planet. A point on the Earth's surface at the equator has a velocity of 463 m s^{-1} relative to an absolute frame of reference such as the position of the stars. By comparison, our nearest planetary neighbours Venus and Mars have absolute velocities on their equators of $1\cdot8$ and 239 m s^{-1}, respectively. The velocities of the atmosphere and ocean relative to the rotating Earth are small and therefore a parcel of air or water will only travel a small distance on the globe compared with the movement of the Earth. Under these circumstances it is to be expected that the rotation of the Earth will have a profound influence on the circulation of both atmosphere and ocean.

First, consider the effect of rotation on a stationary particle on the Earth's surface. The particle will be subjected to the acceleration due to gravity, $\mathbf{g}$, and a centrifugal acceleration in the direction perpendicular to the axis of rotation, as illustrated in figure 7.1. At the point P the outward centrifugal acceleration is $\Omega^2 r_1$, where $r_1 = r_e \cos \phi$. At the equator this acceleration is $0\cdot034 \text{ m s}^{-2}$, compared with an average acceleration due to gravity of $9\cdot81 \text{ m s}^{-2}$. The resultant vector, known as the 'effective gravity', is directed at an angle of $0\cdot1°$ to the true direction of gravity. In both meteorology and oceanography it is assumed that the direction of gravity is that of the resultant gravity vector and therefore the centrifugal acceleration is implicity taken into account. The local vertical is also assumed to be aligned with the resultant gravity vector.

In Chapter 1 it was shown that a geopotential surface is a surface of constant gravitational energy and, by definition, the component of gravity is perpendicular to this surface. Due to the variation of the local component of g with position, it is necessary to define a level surface in terms of a geopotential surface rather than a geometric surface. The geopotential, Φ, is the work per unit mass in moving a parcel of fluid a

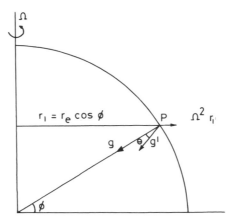

Figure 7.1. Centrifugal acceleration on a particle stationary with respect to the Earth. $\mathbf{g}'$ (effective gravity vector) is the resultant of the centrifugal acceleration and gravity, $\mathbf{g}$.

vertical distance, z, against gravity and is given by:

$$\Phi = \int_0^z g \, dz' \qquad (7.1)$$

where z is the geometric height relative to mean sea level. It is noted that the geopotential has units of energy. The acceleration due to gravity varies horizontally from the pole to the equator by about $0 \cdot 5\%$ and in the vertical by $0 \cdot 23\%$ per 10 km from sea level.

In both oceanography and meteorology a common unit for the geopotential is the dynamic meter. In the atmosphere it is defined as:

$$Z = \Phi/10 \qquad (7.2)$$

and in the ocean as:

$$D = -\Phi/10 \qquad (7.3)$$

The negative sign appears because it is measured downwards from the surface of the ocean.

The geopotential, or dynamic height, is related to the pressure, p, by the hydrostatic equation:

$$\alpha \, dp' = -g \, dz' = -d\Phi \qquad (7.4)$$

For the ocean, the dynamic height at pressure p with reference to sea level in a stationary ocean is:

$$\int_p^0 \alpha \, dp' = - \int_h^0 g \, dz' = - \Phi \tag{7.5}$$

$$\therefore \quad D = \frac{1}{10} \int_p^0 \alpha \, dp' \tag{7.6}$$

It is assumed that the pressure at sea level is zero in the ocean. The geopotential depth, D, can therefore be calculated from numerical integration of the profile of the specific volume, α, obtained from *in situ* measurements of temperature and salinity.

The influence of rotation on a moving particle on the Earth's surface will now be considered and the following example demonstrates how rotation will affect the motion of a projectile, as perceived by an observer on a rotating frame. Consider a turntable rotating with an angular velocity Ω, as shown in figure 7.2. If a person stands at the centre and throws a ball, the ball will take a curved path with respect to an observer on the turntable. It will appear to this observer that the ball has rotated in a direction opposite to that of the turntable. To a stationary observer looking down on the turntable the ball will appear to have been unaffected by the rotation of the turntable and it will be perceived to move in a

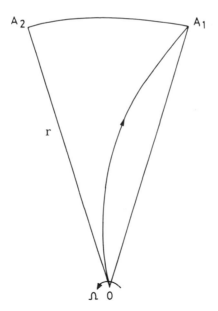

Figure 7.2. Coriolis effect: deflection of a moving particle on a turntable.

straight line. The deflection of the ball is, therefore, the result of the transformation from a non-rotating to a rotating frame of reference. This deflection force is known as the coriolis force.

In figure 7.2 the ball is launched in the direction of OA_1 with an initial velocity v. After a small increment of time, Δt, the ball has reached the edge of the turntable but, because of the rotation of the turntable, point A_1 has rotated to A_2. According to an observer at A_1, the ball has been deflected a distance A_1A_2, perpendicular to the initial direction of the ball. If the radius of the turntable is r then:

$$A_1A_2 = \Omega \; \Delta tr \qquad (7.7)$$

and since $r = v \, \Delta t$, then by substitution in equation 7.7

$$A_1A_2 = v\Omega \; \Delta t^2 \qquad (7.8)$$

A particle under a field of constant acceleration, a, will be displaced by $\frac{1}{2} a \, \Delta t^2$ and therefore by inspection of equation 7.8:

$$a = 2\Omega v \qquad (7.9)$$

First it is noted that the coriolis acceleration, a, is proportional to the velocity in a rotating frame of reference and therefore a stationary body on the Earth will not be subjected to a coriolis acceleration. Second the coriolis acceleration is a vector which acts perpendicular to both the direction of motion of the projectile and to the direction of the axis of rotation, which is vertical in this case. The vector acceleration $\mathbf{a}$, is given by the vector product:

$$\mathbf{a} = 2\mathbf{\Omega} \times \mathbf{v} \qquad (7.10)$$

On a rotating planet the axis of rotation is only aligned in the direction of the vertical axis at the poles, whilst elsewhere it is at an angle of $90 - \phi$, where ϕ is the latitude, as shown in figure 7.1. Therefore, to obtain the coriolis acceleration on a locally horizontally moving particle it is necessary to take the component of rotation in the direction of the local vertical axis. From figure 7.1, this component is $\Omega \sin \phi$ and thus, for horizontal motion, the coriolis acceleration is dependent on the latitude. At the equator the local vertical is perpendicular to the axis of rotation and therefore, for horizontal motion, the coriolis acceleration is absent. It will be noted that, in general, vertical motion will be subjected to a coriolis acceleration as a result of the component in the local horizontal direction. However, because of the small vertical accelerations observed in both atmosphere and ocean and the dominance of gravitational and vertical pressure accelerations, this component of the coriolis force is usually neglected.

For a moving particle, relative to a rotating Earth, an additional centrifugal component will arise. However, because of the relatively low velocity of the atmosphere and ocean, this centrifugal component is small and it is often neglected. For a horizontal motion of 10 m s^{-1} at 45° latitude, the coriolis acceleration is $1 \cdot 03 \times 10^{-3} \text{ m s}^{-2}$, which compares with a centrifugal acceleration of $2 \cdot 21 \times 10^{-5} \text{ m s}^{-2}$. It can therefore be seen that the centrifugal force is only about 2% of the coriolis force. However, both forces are small compared with the acceleration due to gravity.

7.2. Inertial motion

A practical demonstration of the existence of the Earth's rotation can be made by observing the rotation of a Foucault pendulum. Figure 7.3 shows a pendulum oscillating along a line AA' at the North Pole. Assume that there is an observer a small distance from the pole on the Greenwich Meridian. The pendulum will be unaffected by the Earth's position but the observer will be rotated with the Earth. To the observer, the plane of the pendulum's oscillation will apparently rotate in the opposite direction to the rotation of the Earth. After 6 hours the observer will have rotated by 90° and will be aligned with the pendulum and after 12 hours the pendulum will be observed to oscillate in the original direction. This

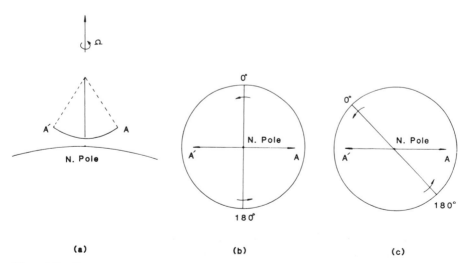

(a) **(b)** **(c)**

Figure 7.3. Foucault pendulum: (a) oscillating at the North Pole; (b) and (c) apparent rotation of the pendulum relative to a rotating Earth.

period of the rotation of the pendulum is known as the pendulum day or the inertial period. At the poles the period is 12 hours but, because it is determined by the component of rotation along the local vertical axis, it will increase as the latitude decreases. At a latitude ϕ, the period of rotation will be $12/sin\ \phi$ hours. At the equator, an east–west oscillating pendulum will be aligned in the same direction as the rotation of the Earth and thus there will be no apparent rotation relative to the Earth. However, if a pendulum oscillated in a north–south plane across the equator it will tend to rotate in an anticlockwise direction in the Northern Hemisphere and in a clockwise direction in the Southern Hemisphere. It will thus produce a figure of eight pattern.

It will now be shown that parcels of fluid on a rotating Earth will exhibit very similar behaviour to that of the pendulum. In a non-rotating system, from Newton's second law, a body will continue in a state of uniform motion unless acted upon by a force. For inertial motion in an absolute frame of reference the rate of change of velocity of the fluid parcel is given by:

$$\frac{D_a}{Dt}(\mathbf{v}) = 0 \qquad (7.11)$$

where the subscript a refers to an absolute frame of reference. In section 7.1 it was shown that there is an additional coriolis acceleration in a rotating frame of reference, where the equivalent relation to equation 7.11 may be written as:

$$\frac{D_r}{Dt}(\mathbf{v}) + 2\mathbf{\Omega} \times \mathbf{v} = 0 \qquad (7.12)$$

where $(D_r/Dt)(\mathbf{v})$ is the acceleration of a fluid element as measured on the rotating Earth.

Equation 7.12 may be written for two horizontal components of motion:

$$\frac{Du}{Dt} - fv = 0 \qquad (7.13)$$

$$\frac{Dv}{Dt} + fu = 0 \qquad (7.14)$$

where u is the velocity in the eastward direction, v is the velocity in the northward direction and $f = 2\Omega \sin \phi$. The subscript r, denoting a rotating frame of reference, has been dropped and, from now on, it will be assumed that all motion is relative to the rotating Earth. The main point to be noted here is that a fluid element, or any other body, cannot continue in a state of uniform motion on a rotating Earth, even if all frictional forces and pressure gradient forces could be eliminated.

Equations 7.13 and 7.14 will now be used to deduce the trajectory of a fluid element initially moving northwards with a velocity v in the Northern Hemisphere where $f > 0$. From equation 7.13 it will be seen that the northward motion will induce an acceleration towards the east, because the coriolis acceleration in the Northern Hemisphere always acts to the right of the direction of motion. The parcel's trajectory will curve towards the east, because of the eastward acceleration, and it will obtain a positive u component of velocity. From equation 7.14 the coriolis force will induce a southwards, or negative, acceleration which will decrease the original northward component. At some stage the parcel will have zero northward velocity and it will then have only an eastward component. After this stage it will continue to accelerate southwards and it will acquire a negative v component of velocity which, from equation 7.13, will induce a westward acceleration in its turn. Thus the parcel will follow a circular trajectory and, providing that the change of latitude of the parcel is very small, it will return to its original position. This circular trajectory is known as an inertia circle.

Eliminating v from equations 7.13 and 7.14 yields:

$$\frac{D^2 u}{Dt^2} + f^2 u = 0 \tag{7.15}$$

This is the equation for simple harmonic motion with frequency f, the coriolis parameter. The period of oscillation is:

$$T_i = \frac{2\pi}{f} \quad \text{or} \quad \frac{\pi}{\Omega \sin \phi} \tag{7.16}$$

It can be shown that the period of the motion is $12/\sin \phi$ hours, as for the Foucault pendulum.

Inertial oscillations are frequently observed in the ocean and they can always be identified by spectral analysis of a current meter time series, as shown in Chapter 6. Figure 7.4 shows an example of an inertial oscillation recorded by a current meter in the Baltic Sea, identified by plotting a progressive vector diagram. The theoretical period of 14 hours 8 min was confirmed by these observations. In the atmosphere, inertial observations are only usually obtained at times when pressure gradients are weak and there is a stable vertical stratification which reduces the effects of surface friction. These conditions tend to occur in anticyclonic flow at night. The oscillations will increase the local wind speed when moving in the direction of the large-scale flow but they will decrease it when moving in the opposite direction. Such oscillations can produce

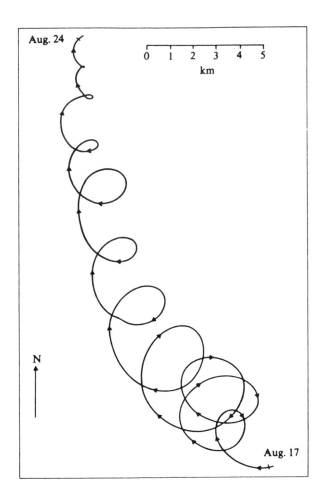

Figure 7.4. *Progressive vector diagram illustrating an inertia current superimposed on a northward motion, observed in the Baltic Sea between 17 and 24 August 1933. (Reproduced from Gustafson, T. and Kullenberg, B. (1936), Sv. Hydr.-Biol. Komm. Skr. Ny. Sev. Hydr., 13, 28 pp.*

local nocturnal jets in the atmospheric boundary layer which have velocities of up to twice the large-scale geostrophic flow. Figure 7.5 shows theoretical examples of inertial flow in the atmosphere whcre the motion is assumed to extend over a large enough range of latitude to produce a significant change in the coriolis parameter, f. It is noted that a variety of patterns across the equator occur, depending on the initial speed and direction of the fluid element. None of the trajectories form closed loops because of the latitudinal variation of f.

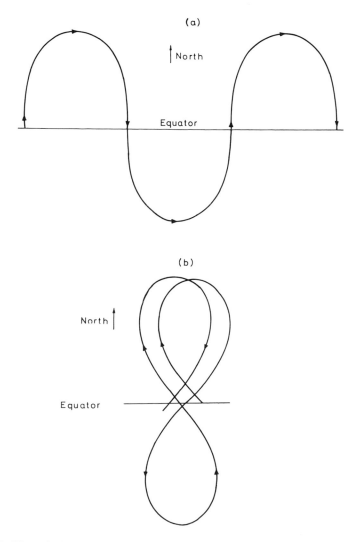

Figure 7.5. Theoretical examples of inertial oscillations about the equator. (a) Initial northward impulse. (b) Initial north-west impulse.

7.3. Pressure gradients and geostrophic motion

The principal force acting in both atmosphere and ocean is the pressure-gradient force. It has already been shown that, in the vertical direction, the dominant balance of forces, known as the hydrostatic balance, is between the upward pressure-gradient force and the downward gravitational force. In the horizontal plane the principal balance of forces is between the coriolis force and the horizontal pressure-gradient force. This balance is known as geostrophic equilibrium.

The horizontal equation of motion, assuming that there is no friction, is now given by:

$$\frac{Du}{Dt} - fv = -\frac{1}{\varrho}\frac{\partial p}{\partial x} \qquad (7.17)$$

$$\frac{Dv}{Dt} + fu = -\frac{1}{\varrho}\frac{\partial p}{\partial y} \qquad (7.18)$$

The right-hand side terms are the horizontal components of the pressure-gradient acceleration and they can be derived in a similar way to that shown in Section 2.4.

For non-accelerating flow on a rotating Earth, Du/Dt and Dv/Dt are both zero and so:

$$-fv_g = -\frac{1}{\varrho}\frac{\partial p}{\partial x} \qquad (7.19)$$

$$+fu_g = -\frac{1}{\varrho}\frac{\partial p}{\partial y} \qquad (7.20)$$

where u_g and v_g are the eastward and northward components of the geostrophic velocity, respectively.

Consider a situation in the Northern Hemisphere where the atmospheric pressure at sea level decreases southwards, as shown in figure 7.6(*a*). The pressure-gradient acceleration will also be directed southwards since $(1/\varrho)(\partial p/\partial y)$ is positive, and, to produce a balance of forces, an equal and opposite directed coriolis force is required acting northwards. Since the coriolis force is perpendicular to the direction of flow (i.e. to the right in the Northern Hemisphere), a westward geostrophic flow will result. In the Southern Hemisphere the coriolis force acts to the left of the motion, therefore a southward pressure gradient will result in an eastward geostrophic flow (figure 7.6(*b*)). In general, clockwise flow occurs around Northern Hemisphere high-pressure and Southern Hemisphere low-pressure systems whilst anticlockwise flow occurs around Northern Hemisphere low-pressure and Southern Hemisphere high-pressure systems. Normally the term anticyclonic flow refers to the flow around a high-pressure centre and cyclonic flow to low-pressure centres. A simple rule for geostrophic motion is that the coriolis force is always directed, following the motion, $90°$ to the right in the Northern Hemisphere and $90°$ to the left in the Southern Hemisphere. As the coriolis force is directed against the pressure gradient, the high-pressure centre must always be in the same direction as the coriolis force.

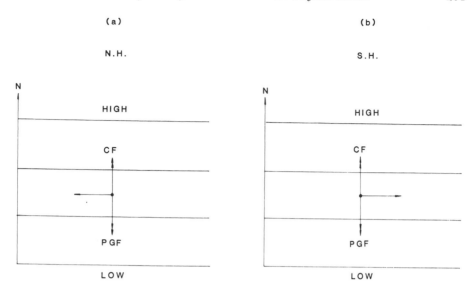

Figure 7.6.   *Balance of forces in geostrophic flow in (a) the Northern Hemisphere and (b) the Southern Hemisphere.*

It must be remembered that geostrophic flow in the atmosphere and in the ocean is only an approximation to the actual horizontal flow. For example, rapid changes in the pressure gradient may occur which would induce inertial accelerations and non-geostrophic flow. In addition, frictional drag of the flow over the Earth's surface will induce deviations from geostrophic flow. On the equator there is no coriolis force and therefore geostrophic flow is impossible. In the atmosphere as a whole, geostrophic flow is a good approximation, being accurate to about 10%, to the observed flow above the atmospheric boundary layer (i.e. approximately 1 km above the Earth's surface) in large- and synoptic-scale systems whose linear dimensions exceed 1000 km. In smaller-scale systems, such as thunderstorms and sea breezes, there are large departures of the winds from geostrophic motion. In the upper troposphere in intense jet streams, where velocities can reach 100 m s^{-1}, the flow is not geostrophic. In the ocean, in contrast to the atmosphere, the geostrophic approximation can be applied to smaller-scale systems, such as mesoscale eddies with a length scale greater than 50 km, as well as to large-scale flows. However, significant departures from geostrophic motion do occur in the surface mixed layer of the ocean, due to local wind-driven currents, and also within 100 m of the bottom, particularly where there are large variations in bottom topography.

Some examples of geostrophic motion in the atmosphere and in the ocean will now be considered. Equations 7.19 and 7.20 can be used directly in conjunction with observations of pressure on a horizontal surface to obtain the geostrophic wind in the atmosphere. Accurate measurements, to within about $0 \cdot 1$ mb or 10 Pa, can be made of the surface pressure using mercury barometers. Horizontal pressure gradients in middle latitudes are of the order of $1-5$ kPa over a distance of 1000 km. Geostrophic winds can therefore be calculated to an accuracy of at least 1%. In practice, pressure observations have to be corrected to mean sea-level pressure ($101 \cdot 3$ kPa), using the hydrostatic equation. This correction may introduce further sources of error because a temperature has to be assumed for the fictitious air mass between the ground height and sea level. Above the Earth's surface, pressure measurements are obtained from aneroid barometers inserted into a radiosonde package. In these expendable packages, the accuracy is only about 1 mb or 100 Pa. Of course, the height of the balloon at a particular pressure has also to be determined before the geostrophic wind equation can be used. From the simultaneous temperature measurements made by the radiosonde package, the hydrostatic equation (equation 7.4) can be numerically integrated from the surface at a known pressure p_0 to the observed pressure p. The geopotential height of each pressure measurement is thus obtained. Another method of deducing the horizontal pressure gradient above the ocean surface, illustrated in figure 7.7, involves making simultaneous measurements of pressure and height from an aircraft. Radar altimeters allow geometric height measurement to an accuracy of $0 \cdot 1$ m but this technique is ultimately limited by deviations of the sea surface from the mean sea level, as a result of horizontal variations in the geoid and of ocean currents.

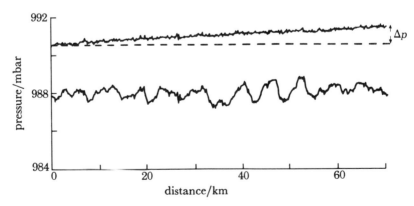

*Figure 7.7. Horizontal pressure gradients measured by an aircraft over the ocean before (lower curve) and after (upper curve) correction for altitude variations. The resulting pressure difference, Δp, can be used in geostrophic wind calculations. (Reproduced from S. Nicholls et al. (1983), Phil. Trans. Roy. Soc., **A308**, 299, figure 3, with permission of the Royal Society.)*

In the ocean there are difficulties in obtaining absolute geostrophic velocities. The first, and most difficult, problem lies in the measurement of the horizontal pressure gradient. In the last 10 years bottom-pressure recorders have been deployed in the Atlantic and Indian Oceans. Their principal objective has been to obtain short-period records of pressure variations for tidal analysis. However, with longer-term measurements, it has been found that there is a drift in the instrument calibration which is of the same order as the longer-period pressure changes. On the ocean bottom one is trying to detect pressure variations between 100 and 4000 Pa, compared with an absolute bottom pressure of about 4×10^7 Pa, over periods of months and possibly years. This implies that the bottom-pressure recorder has to become more sophisticated, and presumably expensive, if this experimental problem is to be successfully overcome. In the meantime, oceanographers have developed an indirect method of obtaining geostrophic currents known as the dynamic method. Unfortunately this method has one important deficiency in that it can only be used to determine relative geostrophic currents. The method is based on the fact that, although it is not practical to obtain an absolute measurement of the pressure gradient at one level in the ocean, it is possible to determine the pressure gradient at one level *relative* to a second level, by use of the hydrostatic equation. The vertical pressure difference between the two levels is dependent on the density, or specific volume, of the sea water which can be determined from temperature and salinity measurements. It has been shown in Section 7.1 that the geopotential difference between two pressure surfaces can be obtained from the integration of the specific volume, α, between the two surfaces. In order to use this information to calculate geostrophic currents, the pressure-gradient acceleration along a constant geopotential surface has to be transformed into the geopotential slope of a constant pressure, or isobaric, surface.

The horizontal pressure-gradient acceleration is given by $(1/\varrho)\delta p/\delta x$. For hydrostatic equilibrium:

$$\delta p = \varrho \, \delta \Phi \tag{7.21}$$

where $g \, \delta z = \delta \Phi$ and therefore:

$$\left(\frac{1}{\varrho}\right) \frac{\delta p}{\delta x} = \frac{\delta \Phi}{\delta x} \tag{7.22}$$

In the limit as $\delta x \to 0$:

$$\frac{1}{\varrho} \left(\frac{\partial p}{\partial x}\right)_z = \left(\frac{\partial \Phi}{\partial x}\right)_p \tag{7.23}$$

where $(\partial \Phi/\partial x)_p$ is the slope of the constant pressure surface, p.

The geostrophic current along a pressure surface p_1 is therefore given by:

$$fv_1 = \left(\frac{\partial \Phi_1}{\partial x}\right)_{p_1} \qquad (7.24)$$

However, it is not possible to determine $\partial \Phi_1/\partial x$ because there is no reference geopotential surface. Although the sea level of a stationary ocean would be a geopotential surface, currents and tidal variations produce significant departures from the geoid. The geostrophic current at a second pressure level, $(p_2 > p_1)$, is

$$fv_2 = \left(\frac{\partial \Phi_2}{\partial x}\right)_{p_2} \qquad (7.25)$$

subtracting equations 7.25 and 7.26

$$v_2 - v_1 = \frac{1}{f}\frac{\partial}{\partial x}(\Phi_2 - \Phi_1) \qquad (7.26)$$

From equations 7.4, the difference in geopotential, $\Phi_2 - \Phi_1$, between the two pressure surfaces is:

$$\Phi_2 - \Phi_1 = -\int_{p_1}^{p_2} \alpha \, dp \qquad (7.27)$$

or alternatively:

$$\Phi_2 - \Phi_1 = -10 \, \Delta D \qquad (7.28)$$

where ΔD is the dynamic height between the two pressure surfaces, measured in dynamic meters. Thus:

$$v_1 - v_2 = \frac{10}{f}\frac{\partial}{\partial x}(\Delta D) \qquad (7.29)$$

A similar equation can be derived for the u component of flow.

Figure 6.10 shows a map of dynamic height, ΔD, between the surface, $p_1 = 0$, and the 1500 db pressure surface (p_2) in the ocean. One decibar pressure is equivalent to 1 dynamic meter which, in turn, is approximately $1 \cdot 02$ m depth. In SI units, 1 db is equal to 10 kPa. The direction of the surface geostrophic flow relative to the flow at 1500 db is indicated on the map. It can be seen that the centres of high dynamic topography correspond to anticyclonic surface flows relative to the 1500 db flow whilst cyclonic surface flows correspond to low dynamic topography. The

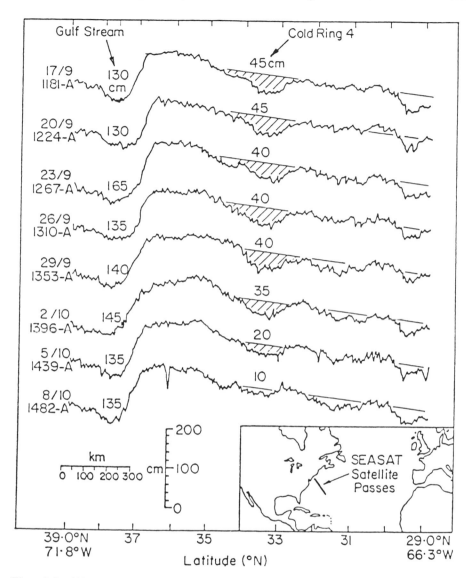

Figure 7.8. Altimeter residuals for eight collinear passes obtained over a 3 week period by SEASAT in 1978 in the Gulf Stream region. (Reproduced from R. E. Cheyney and J. G. Marsh (1981), J. Geophys. Res., 86(C1), 480, figure 9, copyright of the American Geophysical Union.)

resemblance between the dynamic topography maps and the surface current maps has already been noted in Section 6.2. This is related to the fact that, on the large-scale, deep geostrophic currents are usually weaker than surface geostrophic currents and therefore v_1 tends to dominate equation 7.29.

Considerable effort over the last century has been devoted to the independent determination of v_1 or v_2. This can be done by:

(i) Measurement of the current v_2 at a given depth.

(ii) Determination of a depth of no motion, i.e. $v_2 = 0$, from, for example, a tracer experiment.

(iii) Measurement of the total transport through a confined passage using, say, an underwater cable (see Section 6.1). This requires a vertical profile of the relative geostrophic velocity from the surface to the bottom and the profile is obtained by repeated applications of equation 7.29. The geostrophic velocity at each depth can then be adjusted by a constant velocity to give the observed transport over the total depth.

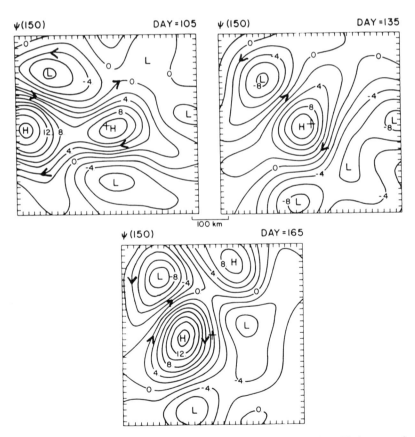

Figure 7.9. The horizontal geostrophic flow at three times, each separated by 30 days at a depth of 150 m. Tick marks on the side indicate the interpolated grid interval (15 km) and the small cross lies in the map centre (69°40'W, 28°N). (Reproduced from J. C. McWilliams (1976), J.P.O., 6, 813, figure 1, with permission of the American Meteorological Society.)

Method (i) can only be used where long-period measurements are available. Method (ii) has been used widely but it often gives contradictory results. In some regions, such as the Antarctic Circumpolar Current, there is an absence of a level of no motion. Method (iii) has been used to determine the geostrophic flow in the Gulf Stream across the Florida Straits but it is not of general use in the open ocean.

A recent development is the satellite radar altimeter which can measure the elevation of the sea surface to an accuracy of at least 10 cm. Figure 7.8 shows the elevation of the sea surface over a north-west to south-east section of the Gulf Stream, measured by SEASAT. The large slope at $37°N$ is the Gulf Stream itself and the dip in elevation at $34°N$ is a Gulf Stream ring, or a cold mesoscale eddy. The sea surface slope across the Gulf Stream is $1·30$ m in 125 km or $1·04 \times 10^{-5}$ which is equivalent to a surface geostrophic velocity of $1·16$ m s^{-1}. This technique may, at last, be able to determine the absolute geopotential slope and provide a reference velocity for geostrophic calculations. However, the method will require accuracies of 1 cm, rather than the present 10 cm, to measure the sea surface slopes in mid-ocean where geostrophic velocities are an order of magnitude less than those of the Gulf Stream.

Another recent technique, used in the MODE experiment, is to use a combination of float information, current meter readings and dynamic heights to obtain an objective analysis of the absolute geostrophic flow. Figure 7.9 shows a sequence of mesoscale eddies over a period of 60 days in a 500 km × 500 km box in the western North Atlantic Ocean. The flow direction can be determined in exactly the same way as for atmospheric geostrophic flow, i.e. anticyclonic flow around the high-presssure areas and cyclonic flow around the low-pressure areas.

7.4. Vorticity and circulation

A useful property of fluid flows in both ocean and atmosphere is the spin, or vorticity, of a fluid element. Vorticity is defined as the spin of an infinitesimal element of fluid about its own axis, as illustrated in figure 7.10. It is a vector quantity and directed along the axis of rotation perpendicular to the surface of the fluid element. Positive vorticity is defined as anticlockwise rotation. In both the ocean and the atmosphere, the large-scale flow is predominantly horizontal because vertical velocities are, in general, much smaller than horizontal velocities. It is therefore the vertical component of vorticity, which arises from the horizontal motion, which is most useful for aiding an understanding of the larger-scale motions.

The vertical component of vorticity, ξ, is defined as

$$\frac{\partial v}{\partial x} - \frac{\partial u}{\partial y}$$

Figure 7.10 shows that vorticity can arise from two distinct flows:

(i) Circular flow.
(ii) Shear flow.

In an absolute frame of reference, in the absence of vorticity sources and sinks, a tube of vorticity is conserved as it moves through the fluid:

$$\frac{D_a\eta}{Dt} = 0 \qquad (7.30)$$

In the rotational frame of reference, the vertical component of absolute vorticity, η, is the sum of the relative vorticity (i.e. the vorticity seen by an observer on the Earth's surface) and the planetary vorticity due to the spin of the Earth. For horizontal fluid motion the planetary vorticity is the component of vorticity in the direction of the local vertical, which is the coriolis parameter, f, where $f = 2\Omega \sin \phi$. Note that the vorticity of solid-body rotation is 2Ω, where Ω is the angular velocity. In the

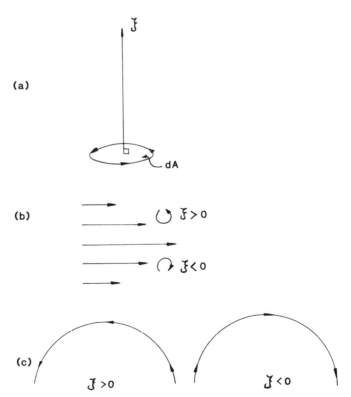

Figure 7.10. (a) Vertical component of vorticity, ξ, in a horizontal flow. (b) Vorticity in a sheared flow. (c) Vorticity in a curved flow.

rotational frame of reference, equation 7.30 becomes

$$\frac{D_r}{Dt}(\xi + f) = 0 \tag{7.31}$$

where ξ is the relative vorticity.

Consider a non-rotational tube of fluid on the equator. The tube has neither relative nor planetary vorticity. However, if the tube is transferred polewards, it must gain planetary vorticity, appropriate to its new latitude. According to equation 7.30 its absolute vorticity, as perceived by a non-rotating observer, must remain zero and so the tube of fluid must have gained a component of relative vorticity of the same magnitude as its planetary vorticity but opposite in sign. In the Northern Hemisphere the coriolis parameter, f, is positive and therefore the acquired relative vorticity is negative and the rotation of the tube of fluid ·is clockwise or anticyclonic. Hence, changes in relative vorticity will be induced by meridional flow due to the conservation of absolute vorticity.

The following two examples will illustrate the application of this conservation principle to flows in the ocean and in the atmosphere. First consider a simple poleward movement of fluid which initially has zero relative vorticity at latitude ϕ_0. As shown in figure 7.11, the fluid will gain relative anticylonic rotation as it progresses polewards in the Northern Hemisphere. The anticyclonic vorticity will induce an anticyclonic curvature in the parcel's trajectory. This will cause the parcel to turn first eastwards, then southwards. As the parcel moves equatorwards it will tend to lose its anticyclonic relative vorticity and, on reaching its original latitude, it will again have zero relative vorticity and its direction will be equatorwards with no curvature. At lower latitudes it will gain cyclonic relative vorticity, in order to maintain the absolute vorticity at its original latitude, and this will, in turn, induce a cyclonic curvature in the trajectory which will eventually rotate the parcel's motion back to a poleward direction. It can therefore be seen that the interchange of relative and planetary vorticity, required to conserve the original absolute vorticity, will produce a horizontal oscillation about the original latitude. These oscillations are known as Rossby waves, or planetary waves, and they can be readily identified in the trajectories of constant-level balloons in the atmosphere (Figure 6.16).

The Lagrangian derivative in equation 7.31 can be expanded into the Eulerian derivative form thus:

$$\frac{\partial \xi}{\partial t} + u\frac{\partial \xi}{\partial x} + v\frac{\partial \xi}{\partial y} + \beta v = 0 \tag{7.32}$$

where $\partial \xi / \partial t$ is the local change in relative vorticity and the second and third terms represent the transport of relative vorticity by the horizontal

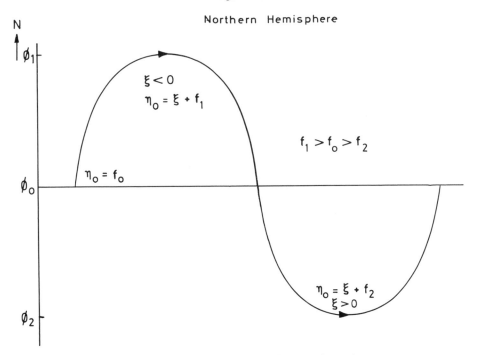

Figure 7.11. *Conservation of absolute vorticity in a Rossby wave (see text).*

flow. $\beta = \partial f/\partial y$ and thus βv is the change in relative vorticity caused by meridional velocity, v.

Consider a simple distribution of meridional velocity given by:

$$v = v_0 \sin(\omega t - kx) \tag{7.33}$$

where ω is the frequency and k is the wavenumber in the x direction. The wavelength is $2\pi/k$. The relative vorticity, ξ, is given by:

$$\xi = \frac{\partial v}{\partial x} = -kv_0 \cos(\omega t - kx) \tag{7.34}$$

Substituting into equation 7.32:

$$+ k\omega v_0 \sin(\omega t - kx) + 0 + 0 + \beta v_0 \sin(\omega t - kx) = 0$$

Thus:

$$\omega k + \beta = 0$$

or

$$\omega = -\beta/k \tag{7.35}$$

The phase velocity is

$$c = \frac{\omega}{k} = -\frac{\beta}{k^2} \tag{7.36}$$

The phase velocity of the Rossby wave is always westwards and it increases in magnitude as the wavenumber of the wave decreases. Table 7.1 shows the phase velocity for different wavelengths at $45°$ latitude. In the middle-latitude atmosphere, the mean tropospheric flow is eastwards and therefore the Rossby waves will move either eastwards or westwards relative to the Earth, depending on the relative magnitudes of the tropospheric flow velocity and the Rossby wave velocity. If the tropospheric flow is greater than the wave speed, the Rossby wave will move eastwards relative to the Earth and if it is less, then the Rossby wave will travel westwards. Very long planetary waves will, therefore, tend to travel westwards and short waves will tend to move eastwards. Should the eastward tropospheric flow match the westward wave speed, the wave will become stationary. The most likely stationary wavelengths lie between 5000 and 10,000 km because their wave speeds correspond to the typical mean wind speeds in the troposphere, as shown in figure 6.5. In the ocean, the large-scale Rossby waves (i.e. those with wavelengths greater than 1000 km) will generally move westwards because any opposing flow in the ocean will be slower than that in the atmosphere.

A second example of the application of the vorticity equation (equation 7.31) yields an explanation of the intense boundary currents which are located at the extreme western edges of the ocean basins and which include the Gulf Stream, the Kuroshio and the Agulhas. First consider the ocean circulation induced by the surface wind distribution. In subtropical latitudes, over all of the major ocean basins, there are semi-permanent, anticyclonic atmospheric flows. This anticyclonic wind stress will also drive an anticyclonic circulation in the upper 500 m of the ocean, by virtue of the vorticity input at the sea surface. The expected response

Table 7.1. Phase speeds and periods of Rossby waves for typical wavelengths, derived from equation 7.26.

Wavelength	Period (days)	Phase speed (m s^{-1})
100	284	$-0\cdot004$
1,000	28	$-0\cdot41$
5,000	$5\cdot64$	$-10\cdot25$
10,000	$2\cdot8$	$-41\cdot00$
20,000	$1\cdot4$	$-164\cdot00$

of the ocean would be a mirror image of the overlying wind system and the maximum dynamic height would be in the centre of the ocean basin, with a broad poleward flow in the western half of the basin and an equatorward flow in the eastern half of the basin. Clearly these expectations do not correspond to reality (figure 6.10) but, if the relative vorticity changes caused by the initial 'mirror image' flow are considered, it will be seen that the poleward flow always gains anticyclonic relative vorticity, and the equatorward flow will always lose anticyclonic relative vorticity. The equatorward flow in the eastern half of the basin will therefore tend to partially negate the anticyclonic vorticity input by the wind system. In the western half of the basin the poleward flow will always enhance the anticyclonic vorticity input of the wind. In time, the anticyclonic vorticity will become more concentrated in the western part of the basin and the gyre circulation will move from its central position towards the west. This will enhance the poleward flow in the western portion of the basin and weaken the equatorward flow in the eastern part of the basin, as illustrated in figure 7.12. Finally, an intense poleward boundary current will form, with very large anticyclonic vorticity, against the western boundary of the basin. The question may be asked—what stops the poleward current from becoming more and more intense, as anticyclonic vorticity continues to be pumped into the ocean? In fact, as more vorticity is added, the intense western boundary current becomes dynamically unstable and energetic mesoscale eddies form. These eddies eventually move out of the intense system and dissipate the energy and

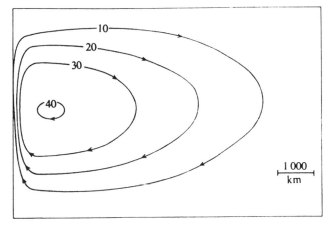

Figure 7.12. The westward intensification of wind-driven currents in a rectangular ocean of constant depth. Streamlines show volume transport in 10^6 m^3 s^{-1}. (Reproduced from The Gulf Stream, *by H. Stommel, p. 92, figure 58, by permission of Cambridge University Press/University of California Press.)*

vorticity of the western boundary current. A number of major, national research programmes are devoted to furthering the understanding of the role of mesoscale eddies in these western boundary currents.

It can also be shown that an equatorward-flowing western boundary current will form for a cyclonic wind circulation. In this case, the pole-ward flow on the eastern part of the wind-driven gyre will gain anti-cyclonic vorticity and it will therefore tend to oppose the cyclonic vorticity input by the wind. At the same time, equatorward flow on the western flank of the gyre will gain vorticity of the same sign as that of the wind. The narrow, southward Labrador current, on the western boundary of the North Atlantic sub-polar gyre, is an important example of an equatorward-flowing western boundary current. This intense, cold current is responsible for the rapid movement of icebergs from the Greenland glaciers into one of the major shipping lanes between North America and northern Europe.

The vorticity equation (equation 7.31) is only applicable to homogeneous, horizontal flows of constant depth. However, in both the atmosphere and the ocean, vertical motion can have a significant in-fluence on the pattern of vorticity and therefore on the circulation. First, consider a simple example in which a cylinder of fluid is rotating with an angular velocity Ω. If the cylinder is now stretched along the vertical axis the diameter of the cylinder will decrease. This will reduce the mo-ment of inertia of the cylinder and, to conserve angular momentum, the cylinder must increase its angular velocity. An analogous situation occurs when a vertical tube of vorticity, having an absolute vorticity η, is stretch-ed along its vertical axis by the vertical velocity gradient $(\partial w/\partial z)$, as shown in figure 7.13. This can be written as:

$$\frac{D\eta}{Dt} = \eta \frac{\partial w}{\partial z} \tag{7.37}$$

It is noted that stretching, $\partial w/\partial z > 0$, will always increase the original vorticity whilst compression, represented by $\partial w/\partial z < 0$, will reduce it. In the absence of vorticity, i.e. $\eta = 0$, vertical motion can neither generate nor destroy vorticity. On a rotating Earth it has been shown that $\eta = \xi + f$ and hence:

$$\frac{D}{Dt}(\xi + f) = (\xi + f)\frac{\partial w}{\partial z} \tag{7.38}$$

An interesting aspect of this vorticity equation is that relative vorticity can be produced by vertical stretching or compression even in the absence of an original relative vorticity because of the vorticity of the rotating Earth.

Consider first an example from the atmosphere. Figure 7.13 shows a

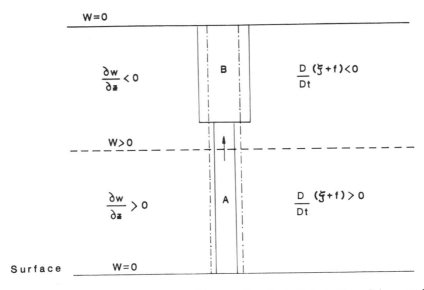

Figure 7.13. Absolute vorticity changes caused by upward motion in the troposphere. It is assumed that the tropopause acts as a lid to the vertical motion.

simple upward motion field, induced by surface warming. If the upper troposphere acts as a 'lid' to upward motion (a simplistic assumption), then it can be seen that stretching of the lower troposphere and compression of the upper troposphere would occur. If it is assumed that there is no original relative vorticity (i.e. $\xi = 0$), then there will be an increase of positive absolute vorticity in the lower layer and an increase of negative absolute vorticity in the upper layer of the troposphere. For small synoptic-scale motions, the change in f with latitude can be ignored and therefore an increase in cyclonic vorticity in the lower layer is expected. Similarly, an increase in anticyclonic vorticity in the upper troposphere may also be anticipated. This type of behaviour is observed over continental regions in summer when surface heating induces a surface cyclonic circulation. Furthermore, the upward motion implies that air near the surface will converge into the circulation and, if this air is very moist, latent heat will be released on ascent. This will increase the buoyancy of the air and it will intensify the vertical and horizontal motions. Such behaviour occurs in tropical cyclones over warm seas where very moist air converges into the cyclone and aids its development. Although other factors are responsible for initializing the tropical cyclone, this effect becomes dominant in its subsequent growth. It is noted that a sinking motion in the troposphere will have the opposite effect and will produce an anticyclonic flow in the lower layer and a cyclonic flow in the upper layer. It can therefore be seen that vertical motion can generate both cyclonic and anticyclonic motions in the atmosphere.

On the larger, planetary scale, the change of the coriolis parameter with latitude becomes an important term in the vorticity balance. In a steady state, where there is no growth of circulation, and the relative vorticity is smaller than the coriolis parameter, then equation 7.38 can be reduced to:

$$\beta v = f \frac{\partial w}{\partial z} \qquad (7.39)$$

This equation says that the relative vorticity produced by vertical stretching is balanced by the relative vorticity lost by meridional motion. Hence, for steady poleward motion, vertical stretching of the lower troposphere will occur and for equatorward motion, vertical compression will take place. Therefore, large-scale poleward movement of warm air will be associated with rising motion and the equatorward motion of cold air will be associated with descending motion. This simple picture goes some way towards explaining the patterns of vertical motion seen in large-scale Rossby waves, usually called baroclinic Rossby waves. Regions of upward motion occur on the upwind side of a pressure ridge and they are associated with the development of surface cyclonic circulations. Surface anticyclones tend to develop on the downwind side of the ridge where downward motion occurs.

Equation 7.39 has been applied in the ocean by Arons and Stommel (1960) to deduce the deep-water flow. It has been remarked earlier that regions of deep sinking motion in the ocean are very limited in horizontal extent and therefore, to balance this local sinking, upward motion is required over the remainder of the ocean. The observation of a permanent thermocline is evidence of the vertical motion of cold water opposing the downward mixing of warmer water. Hence, in the abyssal layers of the ocean, $\partial w / \partial z$ is positive and from equation 7.39, poleward motion is expected. This result ran contrary to earlier ideas of deep-water circulation which had suggested that cold, dense water, formed by sinking in high latitudes, would spread uniformly equatorwards over the width of the ocean basin. Arons and Stommel reasoned that, if cold water moved polewards in the interior of the ocean as predicted by equation 7.39, then it must move equatorwards in a narrow deep western boundary current.

In such a narrow, intense flow, equation 7.39 would no longer be valid because the effects of variations in bottom topography on the vorticity balance would become important. Stommel produced a theoretical model of the abyssal circulation of the world's ocean, driven by sources of deep water in the North Atlantic Ocean and the Weddell Sea, and this is shown in figure 7.14. Subsequent evidence from all of the ocean basins

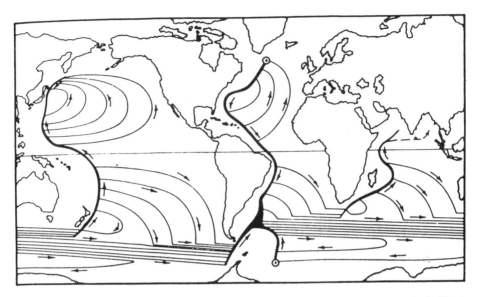

Figure 7.14. A schematic of the abyssal circulation of the world's ocean, driven by sources in the North and South Atlantic. If the two sources, ⊕, in the North and South Atlantic are taken as $20 \times 10^6 \, m^3 \, s^{-1}$ each, then transport between the streamlines is of the order of $5 \times 10^6 \, m^3 \, s^{-1}$. (Reproduced from H. Stommel (1958), Deep Sea Research, **5**, *81, figure 1, with permission of Pergamon Press.)*

has shown that these deep western boundary currents do exist. Their velocities are, however, modest when compared with surface western boundary currents. They are generally less than $10 \, \mathrm{cm\,s}^{-1}$ and the widths of the currents are generally between about 100 and 200 km. There are regions where the circulation shown in this figure is misleading. In the South Atlantic Ocean, the North Atlantic deep water moves southwards along the western boundary but this southward flow overrides the colder and more dense Antarctic bottom water which flows northwards along the same western boundary from the Weddell Sea. The bottom topography also influences the direction of the flow to a marked extent and the Antarctic bottom water, near the equator, moves from the western boundary eastwards to the mid-Atlantic ridge. There is also evidence of potentially cold water in the eastern basin of the Atlantic Ocean, near the equator, which may be derived from Antarctic bottom water flowing through a gap in the mid-Atlantic ridge. Recent studies, using floats and deep-current moorings, below the Gulf Stream have demonstrated that a deep southward boundary current exists in this region but its flow is complicated by a series of permanent eddy circulations. These eddies may be driven by the highly unstable flow in the Gulf Stream but this has not yet been proven.

7.5. *The atmosphere and ocean boundary layers*

In the previous sections the flows of both the ocean and the atmosphere have been discussed in terms of the geostrophic equation. However, close to the interface between the atmosphere and the surface of the globe the flow becomes turbulent. The predominantly horizontal flow becomes broken up into vertical eddies which carry momentum down from the free atmosphere towards the surface. These eddies produce a tangential stress on the land or the ocean when close to the surface. This stress can be seen on the sea surface when a gust of wind produces a ripple pattern, known as cats paws. The stress on the surface, τ, is given by:

$$\boldsymbol{\tau} = \varrho_A C_D (u^2 + v^2)^{\frac{1}{2}} \mathbf{v} \qquad (7.40)$$

where $\mathbf{v}$ is the mean horizontal wind velocity measured 10 m above the surface. ϱ_A is the density of air and C_D is a drag coefficient.

For neutral stability over a sea surface, $C_D = 1 \cdot 5 \times 10^{-3}$, but if there is strong vertical convection, the drag coefficient must be increased. A wind of 10 m s^{-1} will provide a stress of $0 \cdot 1 \text{ N m}^{-2}$. Over land, the drag coefficient varies according to the roughness of the surface. The surface stress on the ocean produces a surface wave field which, in turn, produces vertical motions which mix the surface momentum down into the ocean. These turbulent regions of atmosphere and ocean are known as the boundary layers and they extend to a height of about 1000 m above the Earth's surface and to a depth of approximately 50 m in the ocean. In this section the effects of the boundary layers on flow will be considered. The two boundary layers are different because, in the atmosphere, the boundary layer is removing momentum from the flow and it is therefore acting to retard the flow of the free atmosphere. However, in the ocean the boundary layer is the main source of momentum to drive not only the surface circulation but also the deeper geostrophic circulation.

First, consider the influence of a frictional force on the geostrophic wind. Figure 7.15 shows the balance of forces between the pressure-gradient force, the coriolis force and a frictional force. Basically the frictional force opposes the motion and reduces the wind to below its geostrophic value. The coriolis force is reduced and, in order to balance the pressure gradient, the wind must rotate towards low pressure to allow a component of the frictional force to oppose the pressure gradient. The angle of rotation is typically $30°$ over a land surface and $10-15°$ over the ocean. A component of the flow is directed down the pressure gradient and therefore mass is transferred between high- and low-pressure systems in these frictional boundary layers. By this process friction acts to reduce the horizontal pressure gradients and to spin down circulation systems in the atmosphere. An extra-tropical cyclone will take about 6 days to

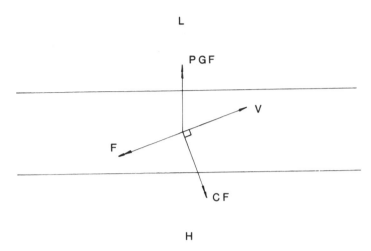

Figure 7.15. Balance of forces in the atmospheric boundary layer. PGF represents the pressure gradient force, CF the Coriolis force and F the frictional force.

dissipate its kinetic energy through the surface boundary layer. It is also noted that the frictional layer will, in turn, produce surface convergence and upward motion in a low-pressure system and surface divergence and downward motion in a high-pressure system. These circulations are known as secondary circulations. However, it must be pointed out that vertical motion in large-scale and synoptic systems is predominantly induced by the flow in the upper troposphere rather than by frictional boundary layers.

   A second example of the effect of friction on motion is that in the surface ocean boundary layer. Consider a region where the horizontal pressure gradient is small and a balance of forces between the surface wind stress, the coriolis force and a frictional force below the surface can be achieved. Here the wind stress will intially drive a current downwind but, because of the coriolis force, which acts perpendicular to the direction of motion, there will be a clockwise rotation of the flow in the Northern Hemisphere (and an anticlockwise rotation in the Southern Hemisphere). For steady flow, frictional drag by the water below the surface layer is necessary to produce a three-way balance of forces. This clockwise rotation of the surface current with respect to the wind stress was first noted by Nansen, when he observed that pack ice drifted at an angle between 20 and 40° to the right of the wind during the Arctic Ocean expedition (1893–1896) on board the research vessel *Fram*. Consideration of the balance of forces below the surface shows that the surface-layer flow will induce a stress on the layer below which, in turn, will produce a further clockwise rotation of the flow. This rotation of the flow with depth was first theorized by Ekman (1905) and it is now known as the Ekman spiral. Below about 50 m these Ekman currents are generally weak. Despite a large number of attempts to observe the Ekman spiral in the ocean, only a handful of measurements have shown its existence and these were taken under pack ice. There are probably two reasons for this difficulty in actually observing Ekman spirals. First, a fluctuating wind would induce strong inertial oscillations in the surface flow. Second, the surface layer is often very well mixed by vertical motion, as the result of surface waves, surface convection and direct wind mixing.

The effect of the ocean surface boundary on the deeper ocean is of crucial importance in the explanation of the climatological current systems above the main thermocline. The Ekman balance between the tangential stress τ, and the coriolis force is given by:

$$-fv = \frac{1}{\varrho} \frac{\partial \tau_x}{\partial z} \qquad (7.41a)$$

and

$$fu = \frac{1}{\varrho} \frac{\partial \tau_y}{\partial z} \qquad (7.41b)$$

Integrating equation 7.41 between the surface and the bottom of the frictional layer, h, yields

$$-\int_{-h}^{0} fv \, \mathrm{d}z = \frac{\tau_x}{\varrho}$$

or

$$-\int_{-h}^{0} \varrho v \, \mathrm{d}z = \frac{\tau_x}{f} \qquad (7.42a)$$

where τ_x and τ_y are the surface components of the wind stress. Similarly from equation 7.41

$$\int_{-h}^{0} fu \, \mathrm{d}z = \frac{\tau_y}{\varrho}$$

or

$$\int_{-h}^{0} \varrho u \, \mathrm{d}z = \frac{\tau_y}{f} \qquad (7.42b)$$

The left-hand sides of equations 7.42a and b are the mass transports in the frictional Ekman layer which are determined by the direction of the stress. The mass transports are perpendicular to the direction of the stress and, in the Northern Hemisphere, they act to the right of the wind stress. Figure 7.16 shows the effect of an anticyclonic and a cyclonic atmospheric circulation on the mass transport in the surface layer. In the anticyclonic case, the mass transport is directed towards the centre of the circulation and, because there can be no net accumulation of mass, mass must be removed from the surface by downward vertical motion. The downwelling, in turn, tends to bow down the thermocline in the central region of the anticyclonic circulation and the accumulation of warm, less dense water will produce a region of high dynamic topography. By the geostrophic equation (equation 7.29), this will cause a surface anticyclonic circulation relative to the deep water. In the cyclonic case surface divergence will cause upwelling of the thermocline and so produce a central region of relatively dense, cold water which will induce a surface cyclonic flow relative to deep water.

Figure 6.11 shows a north–south section of the North Atlantic Ocean. The downwelling region is evident in the sub-tropical zone, as a result of the semi-permanent atmospheric anticyclonic flow, and it is responsible

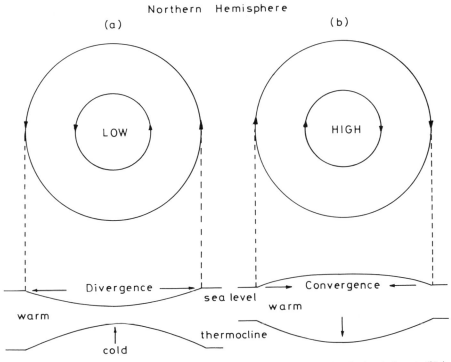

Figure 7.16. Ekman transport induced by cyclonic wind stress (a) and anticyclonic wind stress (b) in the Northern Hemisphere.

for the presence of relatively less dense, warm water in the deeper layers of the sub-tropical gyre. The effect of the sub-polar atmospheric cyclonic circulation is not evident because strong surface cooling and convection tend to weaken the vertical temperature gradients. On a smaller scale, tropical cyclones can have a dramatic influence on sea surface temperatures as the result of strong upwelling induced by the cyclonic circulation in the rear of the storm path.

7.6. *Equatorial winds and currents*

For both the atmosphere and ocean the equatorial region is a dynamically special one. In the atmosphere the winds are both steadier and weaker than the winds at higher latitudes. Synoptic variability of the equatorial atmosphere is less than in middle latitudes, although the seasonal variations associated with the monsoonal circulations are a very significant factor. In the Pacific Ocean interannual variations of the equatorial tropospheric and stratospheric winds are an important feature of its variability. Ocean currents tend to be stronger in equatorial regions than at higher latitudes and they are generally orientated in an east–west direction. They also respond more quickly to wind variations, as seen in the Indian Ocean where current reversals occur in response to the seasonal surface-wind circulation.

These features of the equatorial region can be related to the reduction in the coriolis force close to the equator and to its absence on the equator. At $4°$ latitude, the coriolis parameter $(2\Omega \sin \phi)$ is an order of magnitude less than at $45°$ latitude. The geostrophic and the Ekman balances will therefore no longer be good approximations for winds and currents close to the equator. The question then arises as to how close to the equator can one go before the geostrophic wind equation is of no further use? Generally it is not acceptable to apply the approximation to large-scale atmospheric flows within about $15°$ of the equator and to ocean flows within about $4°$ of the equator, for reasons which are too complex to be discussed in an introductory text. However, the above limitations do not imply that the coriolis acceleration can be neglected. It is rather that other terms in the equations of motion become increasingly important.

Figure 7.17 shows a theoretical calculation of the surface mass transport caused by a steady westward wind stress in a region spanning the equator. On the equator there is no coriolis force and therefore a current transport in the direction of the wind stress is induced. This current will continue to accelerate until a frictional balance between the wind stress and the bottom-layer stress is achieved. Within $1°$ of the equator the coriolis force begins to have an effect and it turns the current to the right in the Northern Hemisphere and to the left in the Southern

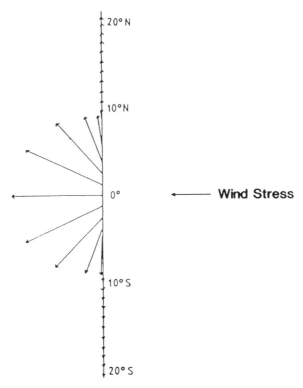

Figure 7.17. Surface transport in the equatorial ocean, caused by a westward wind stress. (Reproduced from M. A. Rowe, PhD thesis, University of Southampton.)

Hemisphere. Therefore there is a divergence of surface flow from the equator which, by the conservation of mass, must be balanced by an upward motion of water into the surface layer. This upwelling is of the order of 1 m day^{-1} and it causes a lifting of the thermocline. The equatorial thermocline usually lies between 100 and 200 m depth but, in local areas where upwelling is strong, it may be less than 50 m below the surface. At about 10° latitude the surface transport is nearly perpendicular to the wind stress, in agreement with the Ekman balance (equation 7.42). In regions where the surface-wind stress is directed eastward such as in the northern Indian Ocean during the south-west monsoon, then the surface transport will converge onto the equator. In this case downward motion out of the surface layer will occur and this will depress the equatorial thermocline.

The meridional tilt of the thermocline, caused by equatorial upwelling, will induce a meridional pressure gradient with high dynamic height polewards of the equator. A few degrees of latitude from the equator, the flow will be close to geostrophic and therefore a westward flow within the thermocline will occur. It is this meridional pressure gradient which is responsible for the north and south equatorial currents which are found in all ocean basins, except for the northern Indian Ocean during the south-west monsoon. In this latter case, because of downwelling on the equator, the meridional pressure gradient is reversed and eastward equatorial currents occur. A downwelling region in the eastern equatorial Pacific and Atlantic Oceans, near $10°N$, results in a similar reversal of the pressure gradient and induces a narrow geostrophic countercurrent towards the east. These examples show the intimate relationship between the surface wind, upwelling and pressure gradients, and the production of strong, zonal equatorial currents.

Along the equator the primary balance of forces in the upper ocean is between the pressure-gradient force and the surface-wind stress. Figure 7.18(a) shows the zonal pressure gradient on the equator in the Atlantic Ocean for February and September. In February the surface winds are generally weak, westward winds and there is only a small eastward pressure gradient. However, in September, very strong westward trade winds extend across the equator and a large eastward pressure gradient is required for balance. The highest dynamic heights occur in the west and here the thermocline is depressed, relative to the eastern part of the region. Figure 7.18(b) shows the depth of the $23°C$ isotherm, which is indicative of the depth of the thermocline and which changes its slope in response to the surface-wind stress. The upward slope of the thermocline towards the east accounts for the frequent appearance of cold, upwelling water in the eastern equatorial Atlantic and Pacific Oceans. In the Indian Ocean, during the south-west monsoon, the zonal tilt of the thermocline is reversed.

If the horizontal pressure gradient exactly balanced the surface-wind stress, there would be no motion along the equator. However, the balance of forces is not exact and the currents respond quickly to the imbalance. Below the surface layer, unimpeded by the wind stress, the eastward pressure gradient in the thermocline drives an eastward undercurrent along the equator. In the Pacific Ocean, the equatorial undercurrent is 14,000 km long but only about 400 km wide. It is confined to within $2°$ of the equator and has a maximum velocity of $2 \mathrm{~m~s}^{-1}$ close to its core in the thermocline. In both the Pacific and Atlantic Oceans the undercurrent slopes upwards towards the east and it can, occasionally, surface in the eastern basins.

The following calculation will give an indication of the maximum eastward acceleration which can occur as the result of the imbalance of the observed pressure gradient with the wind stress. From figure 7.18(a) it can be seen that the maximum difference in dynamic height between $40°W$ and $0°E$ in the Atlantic Ocean is $0\cdot2$ dynamic m. The

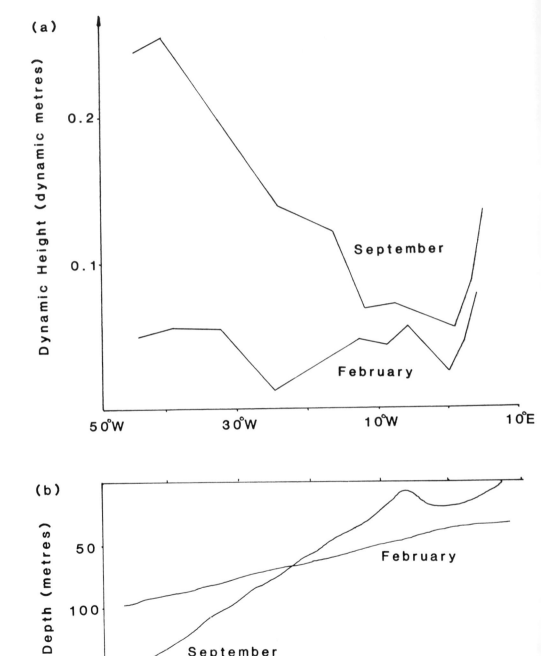

Figure 7.18. (a) Zonal pressure gradient in the equatorial Atlantic in February and September. (b) Depth of 23°C isotherm (equivalent to depth of the thermocline) in February and September.

horizontal distance is $40°$ longitude, or $4\cdot4\times10^{6}$ m. Therefore, the pressure-gradient acceleration is $(0\cdot2\times10 \text{ m}^2\text{ s}^{-2})/(4\cdot4\times10^{6}\text{ m})$ or $4\cdot5\times10^{-7}$ m s^{-2}. The velocity, u, after constant acceleration, a, through a distance, s, is given by $u^2 = 2as$. Assuming uniform acceleration across the Atlantic Ocean:

$$u = \sqrt{(2\times4\cdot5\times10^{-7}\times4\cdot4\times10^{6})} \text{ m s}^{-1}$$
$$= 2\cdot0 \text{ m s}^{-1}$$

This simple calculation therefore gives a reasonable estimate for the maximum observed velocity of the equatorial undercurrent. In practice, frictional turbulence with the deeper ocean and horizontal mixing associated with meanders in the undercurrent will tend to reduce the observed eastward acceleration below that assumed in the above calculation.

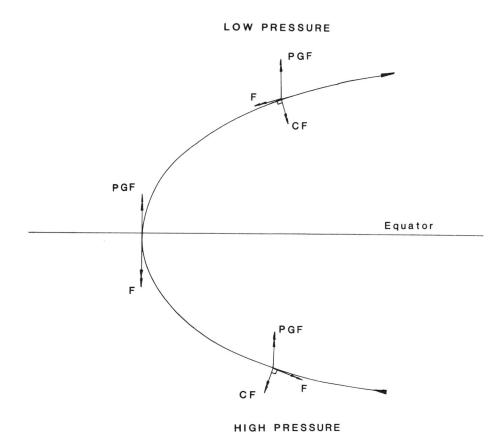

Figure 7.19. Balance of forces in monsoonal flow (summer monsoon) across the equator.

(a)

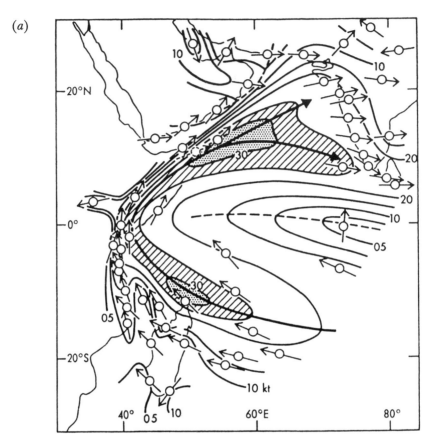

Figure 7.20. (a) East African jet in atmosphere: monthly mean air flow at 1 km in July. Wind speeds at 5 knot intervals. (Reproduced from J. Findlater, Geophysical Memoirs, *No. 115, figure 1, with permission of HMSO (London).) (b) Somali jet in the Indian Ocean. (Reproduced from J. C. Swallow and J. G. Bruce (1966),* Deep Sea Research, **13**, *866, figure 2, with permission of Pergamon Press.)*

Some features of the equatorial atmosphere will now be considered. One of the dominant seasonal variations is the Asian monsoonal circulation which, during the summer, causes a surface atmospheric flow from the southern Indian Ocean to the Asian continent. This flow can be intense, with winds of up to 20 m s^{-1} occurring off Somali. Figure 7.19 is a simplified map of surface pressure for the south-west monsoons and shows that the main feature is the curvature of the winds as they cross the equator. There is a pressure gradient directed from the sub-tropical high pressure in the southern Indian Ocean towards the Indian sub-continent. In the Northern Hemisphere, away from the equator, the geostrophic

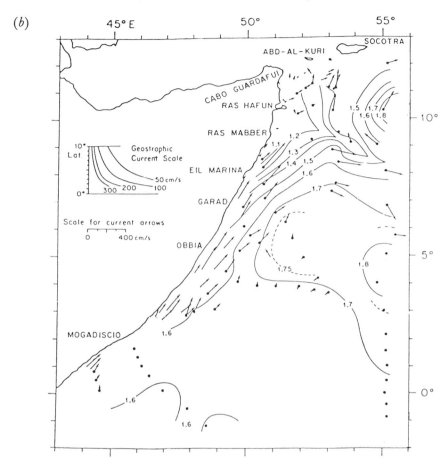

wind is eastwards whilst in the Southern Hemisphere it is westwards. However, because of surface friction, the westward flow will rotate clockwise as shown. Consider a parcel of air in the Southern Hemisphere and moving towards the equator. As the parcel approaches the equator the deflecting coriolis force is weak and the parcel will accelerate north-wards down the pressure gradient. Only friction would be able to reduce the acceleration at this stage. Once the parcel crosses the equator into the Northern Hemisphere it will begin to feel the effect of the coriolis force now acting to the right and it will gradually be deflected towards the east.

A second feature of the monsoonal circulation is the apppearance of an intense, low-level jet stream, as shown in figure 7.20(*a*). It appears off the coast of East Africa during the south-west monsoon. This jet stream, unlike those in the middle latitudes, is directed from the south to the north across the equator and velocities of up to 25 m s^{-1} are reached between a height of 1 and 2 km. It was recently discovered by Findlater, a British meteorologist, who mapped the complete jet stream after receiving reports of very strong, anomalous winds from civil aircraft operating from East Africa. The formation of this jet stream appears to be analogous to that of the western boundary currents in the ocean. In the atmospheric case, the East Africa tablelands provide a convenient western boundary to the low-level monsoon flow. The relationship between the low-level jet stream, the Somali current and the intense upwelling off the Somali coast is a major area of research for both physical oceanographers and meteorologists. In 1979, during the FGGE, a detailed oceanographic and atmospheric study of the summer Indian monsoon was undertaken. Observations showed that the atmospheric jet was exceptionally strong and surface stresses of over $0 \cdot 4 \, \mathrm{N \, m^{-2}}$ were recorded near the Somali coast at $10°N$ in June and July. Sea surface temperatures near the coast dropped from $27°C$ at the beginning of June (i.e. the time of the monsoon onset) to $18°C$ by the end of the month as the result of intense upwelling. Measurements of surface currents by ships' drift in August showed north-eastward currents of between 3 and $4 \, \mathrm{m \, s^{-1}}$ (figure 7.20(b)).

An important feature of the equatorial ocean circulation is its fast response to seasonal variations in wind. In the Atlantic and Indian Oceans, the response time is less than 1 month whilst for the Pacific Ocean, because of its large zonal scale, the response time is between 2 and 3 months. In the Indian Ocean, as discussed above, the Somali current responds to the onset of the south-west monsoon within 2–3 weeks. In contrast, the large sub-tropical gyres in middle latitudes respond on time scales of years to decades and they therefore show little seasonal variation. In this respect, the atmosphere and ocean circulations in low latitudes respond on similar time scales (i.e. seasonal and interannual) and therefore it is in these regions where strongly coupled interactions can take place. This will be discussed in more detail in Chapter 10.

8

Waves and tides

8.1. The spectrum of surface waves

Before discussing the spectrum of surface waves it is necessary to introduce a few definitions. First, the wave height, H, is the vertical distance between the wave trough and the wave crest and it is twice the amplitude, a, of the wave. The wave period, T, is the time interval between the passage of two consecutive wave crests, or troughs, at a fixed point. The wavelength, L, is the distance between two wave crests measured in the direction of propagation. The individual wave speed, c, is given by $c = L/T$, the frequency, σ, by $\sigma = 2\pi/T$ and the wavenumber, k, by $2\pi/L$. The wave, or phase speed, is given by $c = \sigma/k$. The speed of energy propagation of a wave is related to the group velocity of the waves, c_g, where $c_g = \partial\sigma/\partial k$. If the wave speed, c, is not dependent on the wavenumber of the wave, then the wave is said to be non-dispersive and $c = c_g$.

Figure 8.1 shows the energy spectrum of surface waves in the ocean for periods ranging from $0 \cdot 01$ to 10^5 s. The highest-frequency components of the spectrum are the capillary waves which have wavelengths of a few centimetres and periods of less than $0 \cdot 1$ s. They are generated, almost instantaneously, by gusts of wind which produce the familiar 'cats paws' ripple pattern on an otherwise smooth water surface. They are also dissipated in less than 30 s by viscosity. The restoring forces for these capillary waves are, in general, both gravity and surface tension and the wave speed, c, is given by:

$$c = \sqrt{\left(\frac{gL}{2\pi} + \frac{2\pi S}{\varrho L}\right)} \tag{8.1}$$

where S is the surface tension of water, which is typically $74 \times 10^{-3} \text{ N m}^{-1}$. It can be shown that the minimum wave velocity, c_m, occurs when $gL_m/2\pi = 2\pi S/\varrho L_m$, i.e. when $L_m = 2\pi\sqrt{(S/\varrho g)}$, where L_m is the wavelength corresponding to the minimum wave velocity. For sea

Wave period

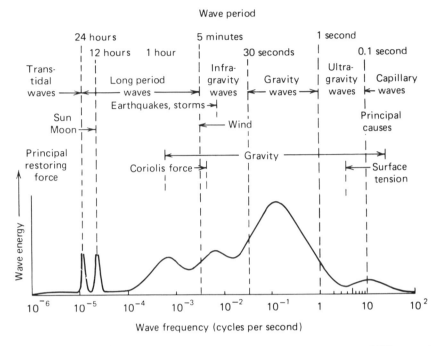

Figure 8.1. Schematic representation of the relative amounts of energy in waves of different periods. (Reproduced from Wind Waves: Their Generation and Propagation on the Ocean Surface, *by B. Kinsman, p. 23, figure 1.2.1, with permission of Dover Publications.)*

water, $L_m = 1 \cdot 7$ cm and $c_m = 23$ cm s^{-1}. At wavelengths shorter than L_m the surface tension in equation 8.1 becomes increasingly dominant and c increases as L decreases. For wavelengths longer than L_m gravity becomes more important than surface tension. In this case, c increases as L increases. An interesting peculiarity of these waves is that the group velocity is greater than the phase velocity and so a packet of capillary waves will move more quickly than the individual waves. The profile of a capillary wave is quite unlike that of a gravity wave in that it has a flat crest and a sharp trough. Capillary waves, because of their dependence on surface tension, are strongly affected by surface films of oil or other organic matter and, because the absence of capillary waves on the surface changes the reflection, surface slicks are easily seen. More recently it has been shown that, because gravity-capillary waves are a sensitive indicator of wind speed, the magnitude of the reflection of microwave radiation ($\lambda = 22$ cm) can give a quantitative measurement of wind speed. The instrument, known as a scatterometer, has been shown to give estimates of wind speed accurate to within 1 or 2 m s^{-1} when flown on the SEASAT satellite in 1978.

The most significant energy band in the spectrum is that of wind waves and swell which have periods between 1 and 30 s. In this frequency band gravity is the only restoring force and therefore these waves are true gravity waves generated by a turbulent surface-wind stress. The particles in these waves execute circular orbits and are not transported by the waves. The radius of the orbits of the particles decreases with depth according to the relationship $r = a \exp[-(2\pi/L)z]$, where a is the amplitude of the wave. At a depth of half the wavelength of the surface wave, the particles' orbit radius is $a/23$. For example, a surface wave having a 1 m amplitude and a 10 m wavelength would have a 4 cm amplitude of particle motion at a depth of 5 m. Therefore wave motions can be seen to be quickly damped with depth. It is noted that the profile of the wave is not exactly sinusoidal. The wave crests are sharper and the wave troughs are flatter than in a true sinusoidal wave. For the larger-amplitude waves this effect becomes more marked, with wave crests reaching a limiting angle of $120°$, corresponding to a wave slope of about 1 in 7. At this stage wave breaking occurs.

For a wave whose amplitude is small compared with wavelength, the wave speed, c, is given by:

$$c^2 = \frac{gL}{2\pi} \tanh\left(\frac{2\pi h}{L}\right) \tag{8.2}$$

where h is the depth of the water. Such waves are known as linear gravity waves and *for waves whose wavelength is smaller than twice the depth of the water* (i.e. $h/L > 0\cdot5$), then $\tanh(2\pi h/L) \simeq 1$ and $c^2 = gL/2\pi$. Since $c = L/T$, it can be easily shown that:

$$c = \frac{g}{2\pi} T \quad \text{and} \quad L = \frac{g}{2\pi} T^2 \tag{8.3}$$

Table 8.1 shows numerical values of c, L and T for typical ocean waves. It is noted that long waves have higher wave velocities than short waves and that the wavelength is proportional to the square of the period. These gravity waves are dispersive since their speed is dependent on wavelength and it would therefore be expected that the group velocity of a packet of wind waves would be different from the phase speed. Since $c = \sigma/k$ and $T = 2\pi/\sigma$ it follows that $\sigma^2 = gk$ or $\sigma = \sqrt{(gk)}$. Differentiating:

$$c_g = \frac{\partial \sigma}{\partial k} = \frac{1}{2}\sqrt{(g/k)} = \frac{1}{2}c \tag{8.4}$$

Therefore the group velocity is equal to half of the wave speed, c. This deduction can be easily verified by studying the propagation of a group of waves produced by a ship or a boat. The individual wave crests will

Table 8.1. Characteristics of wind waves and swell, derived from equation 8.3.

	Wind waves	Swell	Long swell
Period, T (s)	1–5	10	20
Wavelength, L (m)	1·6–39	156	624
Phase velocity, c (m s^{-1})	1·6–7·8	15·6	31·2
Group Velocity C_g (m s^{-1})	0·8–3·9	7·8	15·6

be seen to move towards the observer through the group of higher waves and then they will be seen to die out.

As mentioned earlier, apart from tides, the largest energy density in the surface wave spectrum is associated with wind waves and swell. A simple example illustrates the typical energy in ocean swells. The energy density of a sinusoidal wave is $\varrho g a^2/2$ or $\varrho g H^2/8$ and therefore a wave with a height of 2 m would have an energy density per unit area of 5×10^3 J m^{-2}. If it is assumed that the ocean swell has a 10 s period then its group velocity is $\frac{1}{2} \times 15 \cdot 6$ m s^{-1} or $7 \cdot 8$ m s^{-1}. The energy flux per unit time, impinging on a straight coastline, is the product of the energy density and the group velocity. In the present example, this flux is $7 \cdot 8$ m s$^{-1} \times 5 \times 10^3$ J m^{-2} or approximately 40 kW for each metre of coastline. Most of this energy is dissipated into heat in the wave-breaking zone.

In water depth which is small compared with the wavelength (i.e. $h/L < 0 \cdot 5$), $\tanh(2\pi h/L) \simeq (2\pi h/L)$ and, from equation 8.2, $c = (gh)^{1/2}$. Unlike the ordinary gravity waves for which $h/L > 0 \cdot 5$, the wave speed is independent of wavelength and the group velocity is equal to the phase speed. Such waves are known as long waves and particle motions in long waves are horizontal and do not decay with depth. Long waves appear in the spectrum between swell and tides and they have periods between 30 s and 12 hours.

An extreme example of a long wave is the tsunami, or seismic wave, which can produce locally high-energy densities. These seismic waves are caused by submarine volcanoes, earthquakes or landslides and they propagate quickly, at speeds of about 200 m s^{-1} in deep ocean water (i.e. 4 km depth) and at about 30 m s^{-1} on continental shelves where the ocean is less than 100 m deep. Their wavelengths are very long and, for a tsunami having a wave period of 10^3 s, the wavelength in deep ocean water would be approximately 200 km and therefore the wave would be difficult to detect. However, in shallow water, where it propagates more slowly and where the wave energy is concentrated into a smaller depth, the amplitude of the wave would grow catastrophically.

Other examples of long waves include 'surf beat', which has a period of a few minutes and can be observed as a long-period oscillation in water level on a beach. It is caused by a long wave travelling longitudinally between the surf zone and the shore. A second example is the shelf waves which travel along the 'shelf break' at the edge of the continental shelf. They have periods between 30 min and 2 hours and can travel for hundreds, or even thousands, of kilometres along the shelf break.

The extreme low-frequency long waves are the tides which are forced by gravitational attraction between the Earth, the Moon and the Sun. The tides will be discussed in Sections 8.4. and 8.5.

8.2. Wind waves and swell

In the previous section a number of formulae have been briefly discussed which have been derived from elegant theories of surface waves developed by mathematicians over the last two centuries. It has also been seen that they make available practical tools with which to calculate relationships between wave speed, period and wavelength. However, these formulae and theories give very little insight into how to describe the complicated surface wave patterns observed in nature or into the understanding of how waves are generated and how they decay. It was the need to answer such questions that gave rise to the first quantitative study of ocean waves in the early 1940s. At this time military planners required predictions of wave heights and swell for seaborne operations and, in particular, for beach landings.

During this early period a number of different methods were devised to measure the height of the sea surface covering a frequency range of 1–50 s. One of the most successful instruments was the underwater pressure recorder which could be located either on the bottom, in water less than 100 m in depth, or under the water line of a ship. The pressure fluctuations, a, at depth, h, can be used, in conjunction with wave theory, to give the amplitude of the variation of the sea surface, using the following relationship:

$$a_0 = a_h \cosh(kh) \tag{8.5}$$

where a_0 is the amplitude of the sea surface, k is the wavenumber and a_h is the amplitude at depth h. The signals can be recorded on magnetic tape in the ship or at the shore base. This type of instrument tends to filter out high-frequency fluctuations. A second method involves the measurement of the vertical displacement of a ship or buoy using an accelerometer. Many of the measurements of ocean waves have been obtained from ships equipped with pressure recorders on the hull and accelerometers, the latter being used to correct for the ship's motion. Other methods include the use of inverted echo-sounders, either moored or on the bottom, to obtain surface displacement, and the use of radar and lasers, either on fixed platforms or on aircraft and satellites. All of these methods have their own advantages and disadvantages and, in the end, it depends very much on the question being asked as to what system of measurement is adopted.

A typical trace from a bottom pressure wave recorder is shown in figure 8.2. It confirms previous suspicions that the sea surface is, in general, irregular and does not undergo simple sinusoidal variations. It is clearly not possible to define a single wave height or a single wave period for this record. In order to analyse such a record it is necessary to decompose it into its component frequencies and amplitudes, using Fourier analysis. This is done by dividing the record into blocks of equal periods, say 4 min, and then calculating the Fourier coefficients for each block of the sea-level record, Z_j. If the sampling rate is 2 s, then 120 data points are obtained for each block which, in turn, will give Fourier coefficients, H_n, for 59 discrete frequencies, thus:

$$H_n = \frac{1}{60} \sum_{j=1}^{120} Z_j \exp\left(i\, \frac{2\pi j n}{120}\right) \quad \text{for } n = 1, 59 \tag{8.6}$$

The energy, E_n, for each frequency, n, is then proportional to the modulus of the Fourier coefficients, $|H_n|^2$. In order to obtain a statistically valid sample the procedure is repeated for subsequent blocks of data and the average energy, E_n, is calculated for each frequency. The average energy is then plotted against frequency to produce an energy or power spectrum such as the one shown in figure 8.2. First, this type of analysis shows that, in general, there is a continuous frequency distribution of waves with the maximum energy at the frequency of the dominant wave, F_0. Second, the distribution is not symmetrical in that it has a more rapid reduction towards lower frequencies. Finally, the total energy of the wave field is proportional to the area under the curve.

Two spectral examples will now be considered. Figure 8.3 shows a series of wave spectra observed off the west German coast when the wind was blowing steadily from the land. From common experience one expects the height of the waves, and therefore their energy, to increase with the distance offshore, or the fetch. In the example, the areas under the curves increase markedly with the fetch. The spectra also show that the dominant wave frequency decreases with fetch, from $0 \cdot 4$ to $0 \cdot 2$ cycles s^{-1}. This corresponds to an increase in wave period from $2 \cdot 5$ to 5 s. Secondly, as the fetch increases, the spectra become narrower and this implies that more of the wave energy becomes concentrated in fewer frequencies. At each position the wave spectra are in a steady state. This

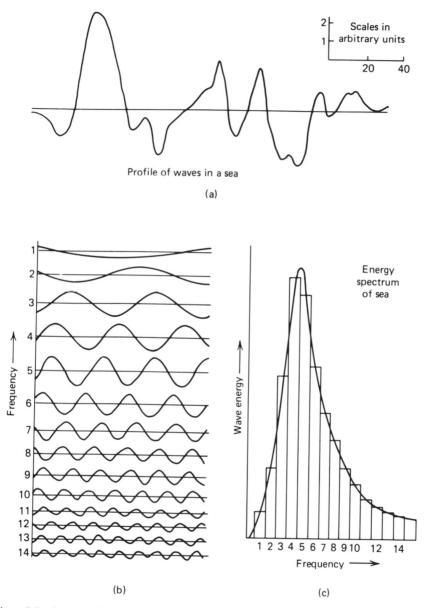

Scales in arbitrary units

Profile of waves in a sea

(a)

Frequency →

Wave energy →

Energy spectrum of sea

1 2 3 4 5 6 7 8 9 10 12 14

Frequency →

(b)

(c)

Figure 8.2. An observed profile of waves in a sea (a). Such a complicated wave pattern can be described as consisting of many different sets of sine waves (b), all of them superimposed. The lower-frequency waves contain more energy than the higher-frequency ones (c). The energy in a wave is proportional to the square of its height. (Reproduced from Oceanography—A View of the Earth *by M. Grant Gross, p. 208, figure 8.4, with permission of Prentice-Hall.)*

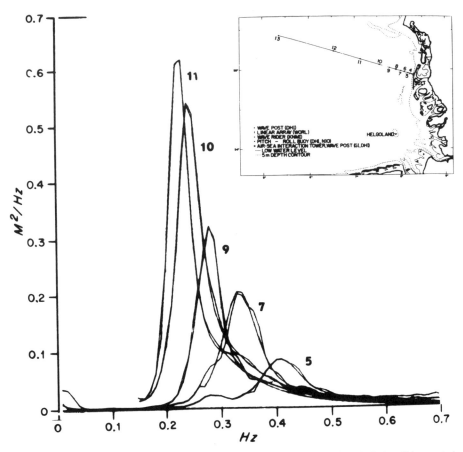

Figure 8.3. Wave spectra of JONSWAP: Evolution of wave spectrum with fetch for offshore winds 11.00–12.00 h, 15 September 1968. Numbers refer to stations inset. (Reproduced from Hasselmann et al. (1973), Dt. Hydrogr. Z. Erg-H (A), **12**, *95 pp., with permission of Deutsche Hydrographisches Institut.)*

means that, although energy is being added to the waves by the wind, energy is being dissipated at about the same rate. Such spectra are known as 'saturated' wave spectra and their shape, dominant wave frequency and total energy can be related to two parameters, namely (*a*) the fetch and (*b*) the wind stress.

The second example is illustrated by figure 8.4 which shows the change of shape of spectra with time under steady wind conditions, i.e. prior to the attainment of a steady state. The high-frequency waves become saturated within a relatively short period of time but the lower-frequency waves continue to grow during the 3 hour period. Unlike the previous example the dominant wave frequency remains relatively con-

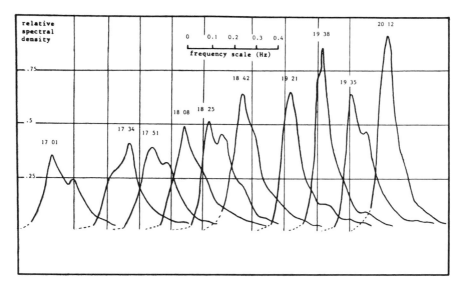

Figure 8.4. Evolution with time of wave spectra. (Reproduced from M. Revault-d'Allonnes and G. Caulliez, p. 110, figure 5, in Marine Turbulence, *edited by J. C. J. Nihoul, with permission of Elsevier.)*

stant during the period, but the spectra exhibit a similar narrowing with time. The time taken to reach a saturated spectrum depends upon the wind speed. For relatively low wind speeds of 5 m s^{-1}, saturation will be attained in about 2–3 hours whilst for higher wind speeds of, say, 15 m s^{-1}, it may take 24 hours. As the fetch increases beyond a critical value, the spectra become independent of fetch and they can be directly related to the wind velocity. At low wind speeds of 5 m s^{-1}, this will occur at a distance of approximately 10 km offshore but for a higher wind speed of 15 m s^{-1} a fetch of more than 500 km is required.

Provided that the wave spectra are in equilibrium with the wind and they are not limited by fetch from the coast, then they can be used to relate the statistics of the wave field to the wind speed. The total energy of the wave field, E_T, at a given wind speed is equal to the area under the spectral curve and the quantity $\sqrt{E_T}$ is a measure of the root-mean-square height of the waves. If the significant wave height, H_s, is defined as the mean wave height of the one-third highest (peak to trough) waves, then it has been shown that:

$$H_s = 4 \cdot 0 \sqrt{E_T}$$

Spectral decomposition is, therefore, a powerful tool for the presentation of large quantities of wave data in a succinct manner and for obtaining empirical formulae which can be used for wave prediction.

Wave spectra also have considerable scientific value in the understanding of how waves are generated and dissipated. The two previous examples show how wave fields grow with time and change with offshore distance, and thus give important clues to the actual mechanisms by which waves are generated. Two mechanisms have been proposed for the development of waves by wind. In the first mechanism it is assumed that a wind blowing across a water surface will produce surface pressure fluctuations as a result of turbulence in the boundary layer. This will, in turn, cause small waves to form. If some of the waves move at a similar speed to these pressure fluctuations, then the pressure field will reinforce the wave field and the waves will grow. Theoretical analyses have shown that the wave energy will grow linearly with time. The important aspect of this theory is the assumption that waves do not affect the pressure fluctuations in the atmosphere and it is, therefore, most applicable to the initial generation of waves when amplitudes are small. The second mechanism assumes that the surface waves will affect the pressure fluctuations in the atmosphere in such a way that surface pressure fluctuations and waves grow with time. If the flow over the waves is smooth then the pressure distribution cannot transmit energy into the waves but, if the flow is turbulent, the pressure distribution can come into phase with the vertical velocity of the sea surface and allow a rapid energy exchange. This causes the waves to grow exponentially with time.

These two mechanisms fail to explain two features of the spectrum; namely the shift of the spectral peak to low frequencies with time and the development of a saturated spectrum. The mechanisms have recently been studied in the JONSWAP experiment in the North Sea. Figure 8.5

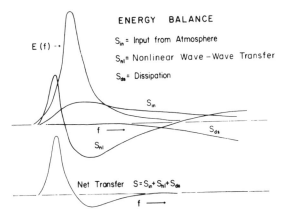

Figure 8.5. Schematic energy balance of the wave spectrum. (Reproduced from Hasselmann et al. (1973), Dt. Hydrogr. Z. Erg-H (A), 12, 95 pp., with permission of Deutsche Hydrographisches Institut.)

shows the wave energy transfer measured for a developing wave spectrum during this experiment. It may be seen that most of the wind energy goes into the spectrum over a broad band of frequencies whilst the dissipation is concentrated at the high-frequency end of the spectrum. In order to achieve a balance between energy sources and sinks, energy has to be transferred to high-frequency waves. This has to be done by interaction between waves of different frequencies. A qualitative example of the process is to consider two sinusoidal waves, with frequencies σ_1 and σ_2, and amplitudes a_1 and a_2. Frequency σ_1 is greater than frequency σ_2. The interaction of these two sine waves is given by $a_1 a_2 \sin \sigma_1 t \sin \sigma_2 t$. Using sum and difference formulae, the following terms are obtained:

$$\cos(\sigma_1 - \sigma_2)t \quad \text{and} \quad \cos(\sigma_1 + \sigma_2)t$$

Therefore the interaction between the two original waves has produced one wave with a higher frequency, $\sigma_1 + \sigma_2$, than either of the original waves and a second wave with a lower frequency. Hence, by this non-linear interaction, energy is transferred to both high and low frequencies. In the wave spectrum the energy which is transferred to the higher frequency is dissipated by wave breaking (white caps and, ultimately, by viscosity). The energy which is transferred to the low frequencies is responsible for the spectral shift to low frequencies during a developing sea, as shown in figure 8.3.

The low-frequency limit of the wave spectrum is the swell which propagates away from the region of wave generation and is dissipated only slowly. The low-frequency swell will propagate with a group velocity greater than that of wind-generated waves and therefore, at a distance from the storm, the lowest-frequency swell waves will be the first to arrive. These will be followed by progressively slower, higher-frequency waves. Consider a wave recording station at a distance L from a storm. From equations 8.3 and 8.4:

$$c_g = \frac{g}{2\sigma}$$

where σ is the wave frequency. The time, t, taken for the swell to travel a distance L is:

$$t = \frac{2\sigma}{g} L \qquad (8.7)$$

Therefore, the frequency of the waves arriving at the recording station will increase with time at a rate proportional to L. From a graph of wave frequency against time, the gradient will give the distance travelled and the intercept will give the time of origin of the waves. If the direction of

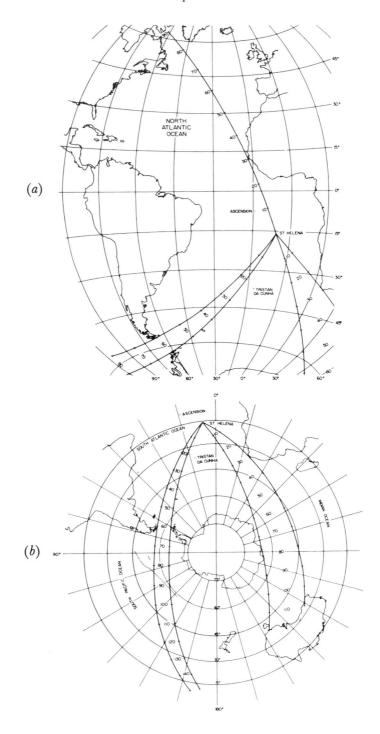

the swell is also measured, or if wave information is available from an additional station, then the position of the storm can be ascertained. Low-frequency swell can travel for very long distances, because of the small attenuation, and the swell will propagate along great circle routes, as shown in figure 8.6. This map shows the possible great circle routes to the Island of St Helena at 16°S. This island has been the subject of much interest in wave research because of the arrival of exceptionally large rollers, or swell, between December and March. From recent work by Cartwright (1977) it has been shown that this swell may originate from winter storms in the Newfoundland region of the Atlantic Ocean. From wave records taken during a large-swell event on St Helena, Cartwright showed that the origin of the swell on the great circle route was coincident with an intense storm in the Newfoundland region 8 days prior to the arrival of the swell. Similar studies have shown swell propagation along great circle paths from the South Pacific to Alaska. It is noted that the group velocity of the swell can be greater than the velocity of the generating storm and so the arrival of swell can give forewarning of a tropical cyclone or a middle-latitude depression. In February 1979 some exceptional swell waves, with a period of 18 s and significant height of 7 m, propagated along the English Channel causing considerable damage to Portland. These waves originated from an intense depression in the central Atlantic.

8.3. Long waves

The main feature that distinguishes long waves from wind waves is that their wavelength is large compared with the water depth and therefore the waves feel the whole depth of the ocean. Water particles oscillate horizontally and are in phase over the depth of the ocean in a long wave. Only close to the bottom boundary will they show any attenuation. Provided that the elevations of the waves are small compared with the mean water depth, then they will propagate with a wave speed $\sqrt{(gh)}$.

Long waves in the ocean may be caused by large falls in surface pressure associated with tropical cyclones or by seismic activity. These waves, like swell waves, tend to follow great circle routes and the location and time of origin of the waves can be obtained by noting the time of their arrival. By this method the origins of major submarine earthquakes have

Figure 8.6. (a) and (b) Great circle routes through St. Helena tangential to the continental land masses. Note that St Helena is exposed to swell from the NW Atlantic and from the Southern Ocean. (Reproduced from D. Cartwright et al. (1977), Q. J. Roy. Met. Soc., 103(438), 664, figures 4a and b, with permission of the Royal Meteorological Society.)

been located in the Pacific Ocean. In deep water such as the Pacific Ocean the variations in bottom depth are relatively small compared with the total depth of the ocean and therefore the wave speed is little affected by the bottom. However, in shallow water the bottom variations reduce the speed of the long waves considerably. As a wave approaches a shore its speed will diminish but its frequency will remain unchanged and so the wavelength becomes shorter and the waves appear to bunch. Mathematically:

$$L = 2\pi \frac{\sqrt{(gh)}}{\sigma} \tag{8.8}$$

The flux of energy per unit time, F, across a coast is the product of the wave-energy density and the group velocity. This flux of energy will remain constant and therefore, because of the reduction in group velocity, the wave-energy density must increase. The wave energy is proportional to the square of the wave height and hence the wave height, H, will increase as the shore is approached. Thus:

$$F \propto \sqrt{(gh)}H^2$$

$$\therefore H \propto \frac{F^{1/2}}{(gh)^{1/4}} \tag{8.9}$$

If a wave has height H_1, in water of depth h_1, then equation 8.9 implies that its height H_2 in water depth h_2 is given by:

$$H_2 = H_1 \left(\frac{h_1}{h_2}\right)^{1/4} \tag{8.10}$$

For a tsunami which has an elevation of 1 m in water 4 km deep propagating into a coastal region having water 10 m deep:

$$H_2 = 4 \cdot 47 \text{ m}$$

Tsunami waves between 3 and 5 m high have been measured in coastal locations in the Pacific Ocean. It is also noted that, for a tsunami wavelength of 100 km in water 4 km deep, equation 8.8 predicts that the wavelength will decrease to 5 km in the shallow water.

For wind waves and swell a similar phenomenon occurs but, because of the increase in slope of the waves, they become unstable and break to form the surf zone. Refraction effects cause the wave front approaching at an angle to the shore to bend towards the shore.

For tsunamis, the wave frequencies are approximately 10^{-3} s^{-1}, or about 15 min. This is considerably less than the coriolis frequency of

10^{-4} s^{-1} and therefore these waves are not affected by the coriolis force. However, for longer wave periods the coriolis force has to be taken into account. Long waves forced by tides or low-frequency meteorological forcing are all considerably influenced by the coriolis force.

Lord Kelvin first described the behaviour of a long wave in a channel, constrained by the coriolis force. Figure 8.7 shows the behaviour of the sea surface and currents in a Kelvin wave. The effect of the coriolis force is to constrain the wave propagation to the boundary of the channel. The elevation, ξ, is given by

$$\xi = \xi_0 \exp(-y/L) \cos(\sigma t - kx) \qquad (8.11)$$

where $L = \sigma/fk = \sqrt{(gh)}/f$, f is the coriolis parameter, σ is the wave frequency, k is the wavenumber and y is the perpendicular distance from the side of the channel.

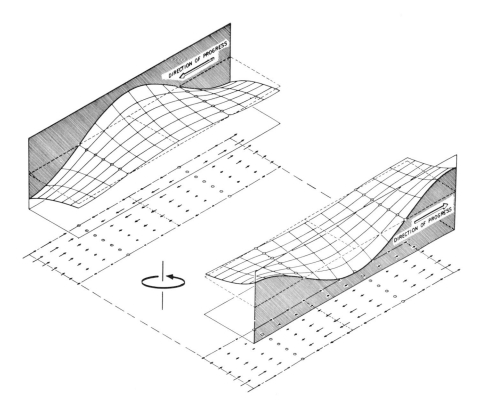

Figure 8.7. Topography of the sea surface for a Kelvin wave in a wide channel in the Northern Hemisphere. (Reproduced from Mortimer, C. H. (1977), Internal waves observed in Lake Ontario during the IFYGL 1972, Spec. Rep. Univ. Wis. Milwaukee, Cent. Great Lakes Studies, 32, 1–22.)

At the side of the channel, $y = 0$ and the wave has a maximum amplitude ξ_0. The amplitude decays with distance and will become negligible for distances much greater than L. For a long-wave speed of 30 m s^{-1}, typical of waves on the continental shelf, $L = 30/10^{-4} \simeq 300$ km. Basically the Kelvin wave propagates as a long wave along the boundary but it is in geostrophic balance normal to the boundary. Therefore the current in the x direction is in geostrophic balance with the pressure gradient caused by the slope in the sea surface, i.e.:

$$fu = -g\frac{\partial \xi}{\partial y} \tag{8.12}$$

Substituting for ξ in equation 8.12 gives:

$$u = \sqrt{(g/h)}\xi \tag{8.13}$$

Therefore high elevation is associated with a positive current and low elevation with a counter current. In a basin closed at one end, a Kelvin wave will propagate around the coast in an anticlockwise direction in the Northern Hemisphere and a clockwise direction in the Southern Hemisphere. Examples of these waves will be discussed with tides and storm surges in later sections of this chapter.

8.4. Internal waves

Consider two fluids having densities ϱ_1 and ϱ_2 in hydrostatic equilibrium. If the interface between the two fluids is moved upwards from the equilibrium position then an element of denser fluid ϱ_2 will be subject to a downward buoyancy force proportional to $(\varrho_2 - \varrho_1)g$. If the two fluids are the ocean and the atmosphere, then ϱ_2 is approximately 1025 kg m^{-3} and ϱ_1 is $1 \cdot 25 \text{ kg m}^{-3}$. In this case the buoyancy force is, to a very good approximation, proportional to $\varrho_2 g$. The fact that the ocean density is so very much larger than the atmospheric density is the reason why the density of the atmosphere is so often neglected in formulae for surface waves. However, for two fluids of similar densities the $(\varrho_2 - \varrho_1)$ term, and hence the buoyancy force, will be small. For example, consider a layer of fresh water overlying a layer of ocean water (a situation commonly encountered in estuaries and fjords), then the difference in density is 25 kg m^{-3}, assuming that the fresh water has a density of 1000 kg m^{-3}. The restoring buoyancy force is only $2 \cdot 5\%$ of the buoyancy force for a surface gravity wave and therefore the restoration of the interface to equilibrium will take considerably longer than for a surface wave. Internal waves, therefore, generally have longer periods and move more slowly than their surface counterparts.

When the layer of lighter water overlies a considerably deeper layer of denser water then the propagation speed of the internal wave is

$$c = \sqrt{\left(\frac{gh_1(\varrho_2 - \varrho_1)}{\varrho_2}\right)} \qquad (8.14)$$

where h_1 is the depth of the upper layer. In a fjord the surface layer may be approximately 4 m deep and therefore, from equation 8.14, the internal wave speed will be 1 m s^{-1}. This may be compared with the surface wave speeds in table 8.1. Figure 8.8 shows the circulation induced by an internal wave in a two-layer fluid. It is noted that the density interface has much larger amplitude displacements than those which occur at the surface and therefore the circulation in the upper layer becomes intensified at the crest of the internal wave in the opposite direction to that of the internal wave. A ship in a fjord will not only produce a surface bow wave but also an internal wave, if the surface layer is shallow. Depending on the speed and length of the boat, the reversed surface circulation at the crest of the internal wave will impede the progress of the boat. This phenomenon is known as 'dead' water because boats caught in these regions will make reduced headway through the water.

A second example of internal waves includes those whose wavelengths are much longer than the total depth of the fluid. Such waves are termed long internal waves. Their wave speed, c, is given by:

$$c = \sqrt{\left(g\,\frac{h_1 h_2}{h_1 + h_2}\,\frac{\varrho_2 - \varrho_1}{\varrho_2}\right)} \qquad (8.15)$$

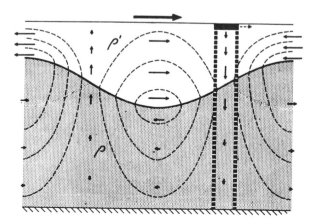

Figure 8.8. Circulations associated with progressive internal waves and their relationship to sea surface slicks (dashes are streamlines).

where h_1 and h_2 are the depths of the top and bottom layer, respectively. The ocean may be considered, albeit very simply, as a two-layer fluid with warm thermocline water, with density $\varrho_1 \simeq 1025$ kg m^{-3}, overlying deep, cold abyssal waters, with density $\varrho_2 \sim 1027$ kg m^{-3}. If $h_1 = 500$ m and $h = 3500$ m, then, from equation 8.15, $c = 2 \cdot 9$ m s^{-1}. This wave speed can be compared with a long surface gravity wave, such as a tsunami travelling at 200 m s^{-1}. For a long internal wave having a period of 1000 s, the wavelength will be $2 \cdot 9$ km which may be compared with a surface wavelength of 200 km.

The amplitude of a typical long internal wave can be estimated in the following way. The hydrostatic pressure variations due to a surface wave of amplitude a_s will be $g \varrho a_s$. The corresponding pressure variations due to an internal wave of amplitude a_I are $g\left[(\varrho_2 - \varrho_1)/\varrho_2) \right] a_I$. If it is assumed that hydrostatic pressure changes will be similar for both waves then:

$$a_I = \frac{\varrho_2}{\varrho_2 - \varrho_1} a_s \qquad (8.16)$$

From the previous example $\varrho_2/(\varrho_2 - \varrho_1)$ is 513 and therefore a long surface wave of height 20 cm may produce an internal wave approximately 102 m in height.

In the above examples it has been assumed that the ocean approximates to a two-layer fluid. In reality, the density field changes continuously with depth and therefore internal waves are rather more complicated than previously suggested. In general, the largest amplitude internal waves are located in regions where the density gradient is a maximum such as the base of the seasonal thermocline or the halocline in the Arctic Ocean. In regions of continuous density variation the buoyancy frequency described by equation 2.9 is a useful parameter. The buoyancy frequency corresponds to the highest frequency of the internal waves that can occur. Also, by analogy with equation 2.9, it can be appreciated that the vertical variation in density will produce corresponding variations in the buoyancy frequency. Hence the internal waves will propagate at different speeds and refraction of the wave fronts will occur. The internal waves will therefore generally propagate both horizontally and vertically. An internal wave generated in the thermocline will tend to propagate downwards into the deeper ocean.

The mechanisms that produce internal waves are numerous. Internal waves in the surface thermocline are often caused by storms but tides and surface currents may also generate them, particularly close to the shelf break or at the entrance to an estuary. They are found literally everywhere in the ocean and often, because of their large amplitudes, cause problems for physical oceanographers. For example, hydrographic data may be contaminated by internal waves which may result in considerable errors in the determination of geostrophic currents by the dynamic method. In addition, sound waves may be refracted by a group of internal waves which will add noise to the received signal.

8.5. Ocean tides

The tides are ubiquitous in the ocean. In the deep ocean tidal currents ranging between 1 and 10 cm s^{-1} are observed in most current meter records and in the trajectories of neutrally buoyant floats. To the dynamical oceanographer these tidal fluctuations represent an annoyance and they have to be removed from the record to obtain long-term currents. Measurements by bottom pressure gauges and, more recently, by satellite altimeters have shown for the first time the regular rise and fall of the tides, typically 10–100 cm in amplitude, over large areas of the deep ocean. Over the continental shelves and especially in shallow, semi-enclosed seas, such as the North Sea, the tidal ranges are an order of magnitude larger than those in the deep ocean. Amplitudes of over 10 m occur in the Severn Estuary, England, whilst in the Bay of Fundy, Nova Scotia, the amplitude is 14 m. The associated tidal currents are usually in the range from 1 to 2 m s^{-1} and are generally larger than wind-induced currents and geostrophic currents. In the English Channel between the island of Alderney and Cap de la Hague tidal currents up to 4·5 m s^{-1} occur whilst near Bodø, north Norway, at the entrance to the Skjerstadt Fjord, currents of 8 m s^{-1} are regularly observed.

The relationship between the tides and the Moon's position has been known since antiquity. Early Roman writers described the twice monthly spring and neap cycle and a Persian philospher, in the second century B.C., showed that the difference between the height of consecutive tides, known as the diurnal inequality, varied with the Moon's position north and south of the equator. The first theory of tides was proposed by Newton (1687) following Galileo's ideas on orbital motion. Newton recognized that the orbital motion of the Moon around the Earth, though itself in equilibrium, would produce an imbalance of forces on the Earth's surface which would, in turn, distort the sea surface.

Figure 8.9 shows the net forces produced by the Moon at different points on the surface of the Earth. If the Earth and the Moon are considered as point masses, then it can be shown that they will rotate about their common centre of gravity with the outward centrifugal force balanced by the central force of gravitational attraction. However, on a sphere, as opposed to a point mass, this balance of forces is only achieved at the centre of mass of the Earth. At a point Z, the zenith on the Earth's surface, the gravitational attraction is greater than that at the centre of the Earth because of its closer proximity to the Moon whilst at the point

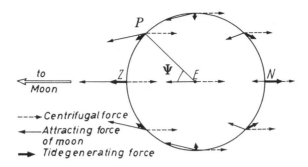

Figure 8.9. Tide-generating force as a resultant of attracting force and centrifugal force along a meridional section through the Earth. Z, N, Moon in zenith or nadir, respectively; E, centre of the Earth. (Reproduced from General Oceanography, *by G. Dietrich, p. 421, figure 180, with permission of Interscience.)*

N, the nadir, the gravitational attraction is less than at the centre of the Earth. Furthermore, because all points on the Earth describe circles of the same radius as the Earth rotates about the common centre of gravity of the Moon and the Earth, the outward centrifugal force at each point on the Earth's surface is the same. Therefore, at the point Z the central gravitational force exceeds the centrifugal force and there is a net force directed towards the Moon. At the point N the centrifugal force exceeds the central gravitational force and there is, therefore, a net force directed away from the Moon.

The net force F per unit mass acting at the point Z is given by the difference between the gravitational force at the centre of the Earth and that at the point Z. The force per unit mass, F_e, at the centre of the Earth is given by:

$$F_e = \frac{Gm_L}{R^2} \tag{8.17}$$

where G is the gravitational constant, m_L is the mass of the Moon and R is the distance between the centres of the Moon and the Earth. Since the distance to the point Z is $R - a$, where a is the radius of the Earth, the net force per unit mass at Z is:

$$F = Gm_L \left[\frac{1}{(R-a)^2} - \frac{1}{R^2} \right] \tag{8.18a}$$

or

$$F = \frac{Gm_L}{R^2} \left[\frac{1}{(1 - a/R)^2} - 1 \right] \tag{8.18b}$$

since $a/R = 1/60$, then by the binomial theorem:

$$\frac{1}{(1 - a/R)^2} = 1 + \frac{2a}{R} + \ldots$$

Hence, neglecting all higher-order terms:

$$F = \frac{Gm_L}{R^3} 2a \qquad (8.19)$$

It is noted that the net tidal force depends on the inverse cube of the Earth–Moon distance whilst the gravitational force depends on the inverse square.

For a general point, P, on the Earth's surface the tidal force can be resolved into vertical (Z_1) and horizontal (H_1) components. It can be shown that these components are given by:

$$Z_1 = \frac{-2Gm_La}{R^3} \tfrac{3}{4}[\cos 2\psi + \tfrac{1}{3}] \qquad (8.20)$$

$$H_1 = \frac{2Gm_La}{R^3} \tfrac{3}{4}[\sin 2\psi] \qquad (8.21)$$

It is noted that ψ will only correspond with the latitude when the Moon is overhead at the equator, as depicted in figure 8.9. Numerical evaluation of Z_1 show that the component is typically 9×10^{-6} of the Earth's acceleration due to gravity. Therefore the tidal force in the direction of the local vertical can be neglected compared with g. Only in the horizontal direction is the tidal-force component comparable with other horizontal forces such as the pressure-gradient force. This is an analogous situation to that described in Section 7.1 where the vertical component of the coriolis force was neglected and the horizontal component retained. The horizontal components of the tidal-generating force, shown in figure 8.9, are directed towards the zenith and nadir. At Z and N, the horizontal component vanishes since $\psi = 0$. This convergence of forces will cause a horizontal acceleration towards the zenith and nadir and, in consequence, a rise in sea level will occur at these two points. A reduction in sea level will occur along the great circle path, at $90°$ longitude to the zenith and nadir. The Moon's orbit varies relative to the Earth's equator and so the latitude of the zenith and nadir will change with time but the relative distribution of tidal forces will remain similar. In addition, it is noted that the rotation of the Earth does not affect the tide-producing forces and therefore, as the Earth rotates, the tide will appear to rotate around the globe with the Moon. The apparent period of rotation of the Moon around the Earth is 24 hours and 50 min and therefore the semi-diurnal tide will be delayed by 50 min each solar day. Hence it will be appreciated that consecutive high waters will be delayed by 25 min.

In addition to the semi-diurnal tide, variations in the Moon's position will produce other longer-period tides. For example, the distance between the Earth and the Moon varies with a period of $27 \cdot 55$ days, and a larger range of tides will occur when the Earth is closest to the Moon. In addition, the angle between the position of the Moon and the Earth's equatorial plane, known as the declination angle, has a period of $27 \cdot 32$ days and this variation also affects the tidal ranges.

The tidal forces due to the Sun can be obtained in exactly the same way as described for the Moon. The ratio of the tidal force produced by the Sun compared with that of the Moon is given by $(m_S/m_L)(r_L{}^3/r_S{}^3)$, from equation 8.19, where m_S and m_L are the masses of the Sun and Moon, respectively, and r_S and r_L are the respective orbital distances from the Earth. The ratio has a value of $0 \cdot 46$ and therefore the lunar tides will always dominate the solar tides. These solar tides will have semi-diurnal and diurnal periods as well as a semi-annual period which is associated with annual variations in the Earth–Sun distance and in the declination angle. Table 8.2 shows the principal tidal frequencies associated with the orbital variations of both the Moon and the Sun relative to the Earth. These tidal frequencies are known as the astronomical tides. Detailed astronomical calculations of the orbits of the Sun, Earth and Moon have shown that 396 distinct tidal frequencies can be identified.

Despite the considerable qualitative success of the astronomical theory of tides, the observations of tides, which commenced in the seventeeth century, show considerable discrepancies with the theory. First, the amplitude of the surface elevation for the semi-diurnal, or M_2, tide is $25 \cos^2 \psi$ cm according to the astronomical theory. For $\psi = 45°$, the theoretical amplitude is $12 \cdot 5$ cm whilst observations of the M_2 tide in the

Table 8.2. Principal astronomical tides.

Tide	Generating force	Type	Period
M_2	Moon	Semi-diurnal	12 hours 25 minutes
S_2	Sun	Semi-diurnal	12 hours 0 minutes
N_2	Moon	Semi-diurnal	12 hours 40 minutes
K_2	Moon and Sun	Semi-diurnal	11 hours 58 minutes
K_1	Moon and Sun	Diurnal	23 hours 56 minutes
O_1	Moon	Diurnal	25 hours 49 minutes
P_1	Sun	Diurnal	24 hours 4 minutes

North Atlantic Ocean reveal amplitudes which are an order of magnitude higher. Only in the Mediterranean Sea is the observed tide of the predicted amplitude. Second, the M_2 high tide at a particular place should, in theory, correspond to the passage of the Moon over the meridian but observations show a delay in the time of high water by as much as a few hours. The magnitude of this delay varies from place to place. Thus, the 'equilibrium theory', though important as a qualitative description of tides, could not be used for tidal prediction.

In the eighteenth and nineteenth century, it was recognized by Laplace and other mathematicians that the tidal wave would be distorted by:

(i) The horizontal component of the coriolis force.
(ii) The distribution of the continents.
(iii) The variation in the depth of water as the tide travels round the globe.

Therefore, for a complete theory of tides, it would be necessary to solve the complete set of hydrodynamic equations on a rotating sphere subject to the astronomical tide-generating force. Although the mathematicians Laplace and Hough did find solutions to these equations, they were too simple to describe the complex tidal response of an ocean basin with complicated bottom topography and irregular coastlines. The gulf between tidal theory and observation prompted the adoption of a different approach to tidal predictions which is still widely used today. This approach is known as harmonic analysis. It had long been recognized that, although the response of the ocean to the astronomical tidal forces was complicated, each unique astronomical frequency would produce a corresponding signal in each tidal record, provided that the ocean behaved as a linear system. Thus, a tidal record could be described by a summation of sinusoidal waves having astronomical frequencies (table 8.2), and, for N frequencies, the sea surface elevation is given by:

$$\xi = \sum_{i=1}^{N} a_i \cos(\sigma_i t - \phi_i) \tag{8.22}$$

where a_i and ϕ_i are the amplitude and phase corresponding to an astronomical frequency σ_i. In order to calculate a_i and ϕ_i it is necessary to have records at least ten tidal periods long and, in practice, because of the spring–neap cycle, it is usual to take a minimum record of 28 days' duration.

Table 8.3 shows the harmonic analysis of tidal records from Immingham (U.K.) and Manila (Philippines). It can be seen that, at Immingham, the semi-diurnal, or M_2, tide is dominant whilst at Manila the diurnal, or K_1, tide is more important. The harmonic analysis shows that, in addition to the dominant components, there are a number of

Table 8.3. Amplitudes of constituent tides at Immingham, U.K., and Manila, Philippines.

	M_2	S_2	N_2	K_2	L_2	O_1	K_1
Immingham (amplitude, cm)	223·2	72·8	44·9	18·3	12·1	16·4	14·6
Manila (amplitude, cm)	20·3	—	—	—	—	28·3	29·7

other frequencies present. At Immingham, the S_2, N_2 and K_2 semi-diurnal tides are present and in all it is possible to distinguish 23 harmonic components. At Manila another diurnal component, O_1, can be seen whilst the only semi-diurnal tide is M_2.

Having obtained the phases and amplitudes of the tides for each astronomical frequency it is then possible to make tidal predictions indefinitely into the future. However, in harmonic analysis, other components appear which do not correspond with astronomical frequencies. First, seasonal variations in wind stress, atmospheric pressure and ocean temperature all cause annual fluctuations in sea level. Second, long-term changes in sea level may be present in a tidal record as the result of climatic variations and relative movement of the land. Thirdly, the hypothesis that the ocean is a linear system is not valid in shallow water; i.e. water less than 100 m deep, where large amplitude tidal currents occur. Non-linear interactions between the tidal components produce higher-frequency harmonics. For example, the M_4 and M_6 tides are harmonics of the M_2 semi-diurnal tide and they have periods of 6 hours 12 min and 3 hours 6 min, respectively. In the English Channel, the M_4 tide has an amplitude as large as the M_2 component and it produces a double high water at a number of ports. Indeed, at Southampton, the M_2 and M_4 interaction gives an 8 hour long stand of high water. Other major harmonics occur between M_2 and S_2 giving frequencies of $M_2 - S_2$ and $2M_2 - S_2$.

From the harmonic analysis it is a straightforward process to obtain the range of spring and neap tides. For a station such as Immingham, where S_2 and M_2 are the dominant constituents, the range of the spring tide, when the tidal forces of the Sun and the Moon are in phase, is given by $2(a_{M_2} + a_{S_2})$. The range of the neap tide at such a station is $2(a_{M_2} - a_{S_2})$ and at this time the Sun and the Moon are $90°$ out of phase. For Immingham (table 8.3), the spring range is 5·92 m and the neap range is 3·00 m.

Harmonic amplitudes and phases for a given frequency can be used to obtain a geographical representation of the tide in an ocean basin or a sea. The lines of constant phase are known as co-tidal lines and they correspond to all points which are at the same tidal stage relative to the Moon's position with respect to the Greenwich meridian. The lines of constant amplitude, or co-range, correspond to points having the same

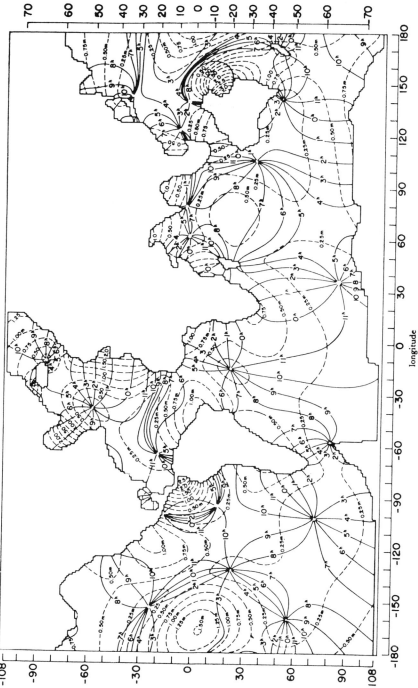

Figure 8.10. Co-tidal map of the modelled M₂ (semi-diurnal lunar) tide. Solid line: co-tidal lines in lunar hours. Dashed line: co-amplitude lines in metres. (Reproduced from C. L. Pekeris and Y. Accad (1969), Phil. Trans. Roy. Soc. A, 265, 433, figure 10, with permission of the Royal Society.)

amplitude. Figure 8.10 shows co-tidal and co-amplitude lines in the world's oceans, predicted by a model, for the M_2 tide. It can be seen that the M_2 tide rotates in an anticlockwise direction about a point in the centre of the Atlantic Ocean, known as the amphidromic point. At this point the tidal range is a minimum. Typical amplitudes in the Atlantic Ocean range from 50 cm in the centre up to 150 cm on the eastern Atlantic coast. The highest tidal amplitudes tend to be found close to the continental coasts. Generally, amphidromic points occur in the central parts of ocean basins and this is why mid-ocean islands often have small tidal ranges. In the North Sea two amphidromic points exist. One is located in the southern North Sea and the other is just off the southern Norwegian coast, and the M_2 tides rotate in an anticlockwise direction around the North Sea. Measurements made in the last 10 years using bottom pressure recorders in the deep ocean have verified the existence of amphidromic points in most of the ocean basins.

As briefly discussed earlier, the equilibrium theory of tides cannot account for the amphidromic points and the observed ranges in sea surface elevation. This arises because it is not possible for the sea surface to come into equilibrium with the tide-generating forces on the short time scale of the major tides. An ocean forced by tidal frequencies will produce long gravity waves whose frequencies will be similar to the coriolis frequency. Such long gravity waves will therefore be considerably modified by the horizontal component of the coriolis force. In a semi-enclosed basin, these waves will propagate as Kelvin waves (described in Section 8.3) and they will be reflected and rotated by the boundary. This rotation will be in an anticlockwise direction in the Northern Hemisphere and this corresponds, in a qualitative manner, to the behaviour of the M_2 tide in the North Atlantic Ocean.

During the nineteenth and twentieth centuries many mathematicians studied the behaviour of long waves in a variety of idealized basins of uniform depth by analytical methods. However, more recently, direct numerical solutions of the linear hydrodynamic equations for the world's oceans, including realistic bottom topography and coastlines, have been obtained by forcing the models with astronomical tidal forces. These models have shown that good agreement with observations can be obtained, as had originally been suggested by Laplace (1778). However for the tides on the continental shelves, there exist non-linear interactions. The variations in bottom topography are also important in these regions and frictional processes play a crucial role. For all the above reasons, more complicated models are required to give reasonable predictions of tidal elevations and currents on the continental shelves.

8.6. Storm surges

In the previous section it was shown that tidal theory can give predictions of sea level to an accuracy of a few centimetres. However, systematic departures from tidal theory predictions do occur as the result of variations in surface pressure and wind stress. Generally these departures are of the order of a few centimetres and they go largely unnoticed. Occasionally, however, they may be of the order of 1 m and, in combination with a high spring tide, can produce considerable flooding of low-lying coastal regions. On 31 January and 1 February 1953, one such storm surge in the North Sea reached a height of between 2 and 4 m above the predicted sea level and inundated large areas of eastern England and Holland, with a loss of 1700 lives. Surges also occur in other parts of the world where there are partially enclosed seas which are relatively shallow. The northern Adriatic Sea, including the Venetian region, is vulnerable to surges caused by intense, winter depressions in the western Mediterranean Sea. Tropical cyclones in the Bay of Bengal have been known to produce surges of between 4 and 7 m along the coast of Bangladesh. These latter surges are particularly devastating because of the low relief of the Ganges delta and they have been known to flood areas up to 100 km inland. Although positive surges in sea level are of importance to those on land, to the master of a ship negative surges can be equally important. For instance, a negative surge of 1–2 m will cause difficulties for a large oil tanker navigating a shallow entrance to a port and, in exceptional cases, it may lead to a ship floundering on a sandbank.

A 1 mb reduction in surface pressure will cause a 1 cm rise in sea level, assuming that the water is in hydrostatic equilibrium. An intense depression, similar to that associated with the 1953 storm surge in the North Sea (figure 8.11), having a central pressure of 970 mb will produce a rise in sea level of approximately 42 cm, assuming a mean surface pressure of 1012 mb. This rise in sea level is about one order of magnitude less than the observed value and it is therefore apparent that most of the surge must be accounted for by the direct effect of wind stress on the sea surface. Consider now a steady wind stress, τ, blowing against a coast. In equilibrium, the wind-stress force must balance the pressure gradient caused by the slope in the sea surface, $\partial \xi / \partial x$,

$$g \frac{\partial \xi}{\partial x} = \frac{\tau}{\varrho h} \qquad (8.23)$$

where h is the depth of the sea. For a given wind stress the slope in the sea level is inversely proportional to the depth of the sea. Thus it is only for shallow seas that a major change in sea level is expected.

A steady wind of $30 \ \mathrm{m \, s^{-1}}$, or 60 knots, will produce a stress of approximately $0 \cdot 9 \ \mathrm{N \, m^{-2}}$, using the bulk aerodynamic formula (equation 5.9). For the North Sea $h \sim 40$ m and therefore from equation 8.23 the sea surface slope is $2 \cdot 2 \times 10^{-6}$. The North Sea is about 600 km in length and therefore the increased height of the sea surface in the southern North Sea, produced by the southward wind stress, would be approximately

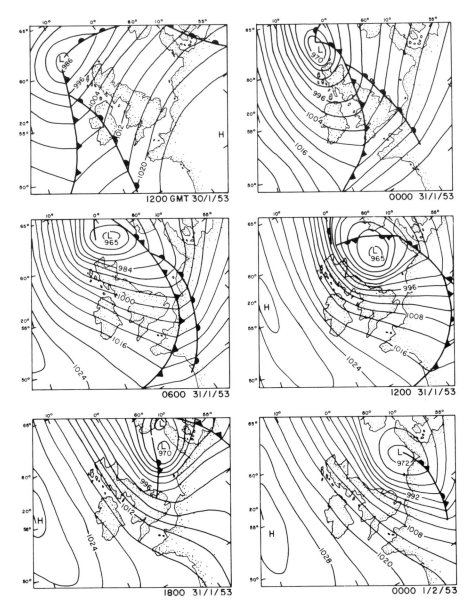

*Figure 8.11. The North Sea floods of 1953: meteorological charts for the period 12.00 hours GMT, 30 January to 00.00 hours GMT, 1 February 1953 (contour interval 4 mb). (Reproduced from R. A. Flather (1984), Q. J. Roy. Met. Soc., **110**(465), 594, figure 1, with permission of the Royal Meteorological Society.)*

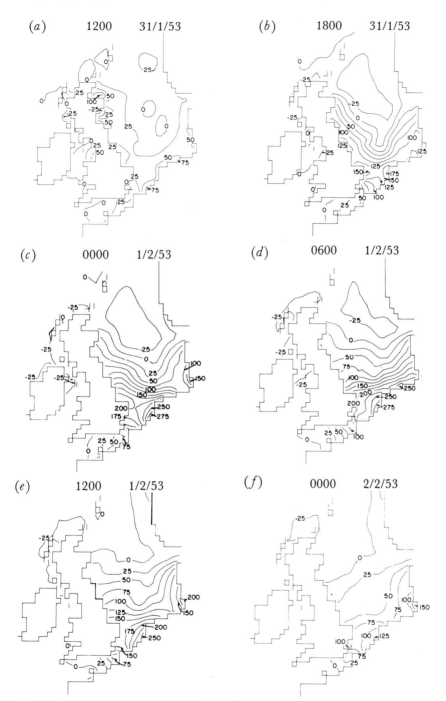

*Figure 8.12. The North Sea floods of 1953: surge elevations computed from a numerical model for the period 12.00 hours GMT, 31 January to 00.00 hours GMT, 2 February (contour interval 25 cm). (Reproduced from R. A. Flather (1984), Q. J. Roy. Met. Soc., **110**(465), 605–608, figure 5, with permission of the Royal Meteorological Society.)*

1·3 m. Hence it can be seen that a surface-wind stress similar to that which produced the 1953 storm surge can account for a much higher sea surface elevation than pressure factors alone. The total estimated rise in sea level due to both wind stress and pressure is approximately 1·7 m, which is lower than that observed along the English and Dutch coast in 1953. The reason for this discrepancy is related to the effect of the focusing of the surge by estuaries, by the very shallow depth of water along the southern North Sea coast and time-dependent dynamics.

An additional feature of North Sea surges is that they propagate in an anticlockwise direction around the coasts. Figure 8.12 shows the surge elevation on 31 January and 1 February 1953 for The North Sea. The surge was first observed at Aberdeen at 14.00 hours on 31 January 1953 with a height of 0·6 m, and subsequently travelled down the east coast and arrived at Chatham 11 hours later, when its height was 2·5 m. It then followed the Belgian and Dutch coasts, where it reached a maximum height of 3·2 m. Its propagation was very similar to that of the semi-diurnal tide which also travels anticlockwise about the North Sea. This surge propagation is related to the coriolis force, which has been neglected in equation 8.23. First the coriolis force causes a rotation of the wind-driven current to the right in the Northern Hemisphere and thus a southward wind stress induces a transport of water towards the eastern coast of England which, in turn, induces a rise in sea level. This rise in sea level is then transmitted along the coast by a Kelvin wave, which is trapped to the coast by the coriolis force. The speed of the Kelvin wave for the North Sea, assuming a depth of 40 m, is calculated to be 20 m s^{-1} (in an anticlockwise direction). The coastal distance between Aberdeen and Chatham is 800 km and therefore the expected time delay, for the theoretical speed of 20 m s^{-1}, is 11·1 hours which is remarkably close to the observed value. Thus it has been shown that wind stress, surface pressure and the coriolis force can account for many features of North Sea storm surges.

The southward Kelvin wave propagation makes it possible to predict storm surges on the east coast of England some 6–12 hours in advance from the observation of tidal residuals (i.e. the difference between the actual sea level and the sea level predicted from tide tables) on the northeast coast of Scotland. This technique has been used routinely for sea-level prediction by the Meteorological Office for the last two decades. However, numerical models using the 24 hour forecasted winds are now being used to obtain accurate predictions for all parts of the North Sea. Similar techniques have also been used in the Adriatic Sea to give predictions of sea level for up to 12 hours ahead. All forecasts are limited by the predictability of the position and intensity of the storm in question. For regions such as the Bay of Bengal accurate sea-level predictions will require detailed forecasts of the paths of tropical cyclones.

9

Energy transfer in the ocean–atmosphere system

9.1. Modes of energy in the ocean–atmosphere system

Table 9.1 shows the magnitude of the major energy sources and sinks in the ocean–atmosphere system. It shows that the major fraction of the incident solar energy is used in the evaporation of water into the atmosphere whilst the kinetic energy of the winds, dissipated into heat by friction, is a relatively minor fraction of the solar flux. However, even the dissipation of the winds' kinetic energy is at least one order of magnitude larger than the other energy sources listed. The direct conversion of visible solar radiation into carbohydrate by photosynthesis is seen to be a relatively inefficient process when compared with the total flux of radiation. Its estimated value, which is based on values of photosynthesis on land, is probably too large because primary productivity in the oceans is limited by a number of factors, such as the supply of nutrients. Most of the energy stored by photosynthesis will be returned to the ocean or atmosphere by oxidation of organic matter, either by burning or by decomposition, and only a small fraction will be laid down in sediments as potential fossil fuels. It is interesting to note that world energy production by the burning of fossil fuels, though small, is now a non-negligible fraction of photosynthetic production. It can be seen from table 9.1 that the world energy production would, however, have to increase one-hundredfold before it matched the dissipation of kinetic energy by the winds. Geothermal heat, produced by radioactive decay in the earth's interior, is very small on the world-wide scale even though it can be locally large in volcanic regions. Dissipation of energy by ocean tides is virtually insignificant on a global scale. However, the majority of the dissipation occurs on the continental shelf, especially in the Bering Sea, the Sea of Okhotsk, the Irish Sea and the Patagonian Shelf where energy fluxes are of the order of 10^3 times larger than the world-wide value.

In the discussion of the energy budget of the ocean and atmosphere it is, therefore, possible to neglect all of the sources of heat except the

259

Table 9.1. Magnitude of major energy sources and sinks in the Earth–ocean–atmosphere system (watts per square metre).

	Energy flux (W m^{-2})
Solar radiation	340
Latent heat	70
Rate of kinetic energy dissipation	~2
Photosynthesis	~0·1
Geothermal heat flux	0·06
World energy production	0·02
Solar reflection from full moon	0·014
Ocean tides	~0·006

incident flux of solar energy. Reference to figure 3.11 will show that, apart from the 30% of the incident solar flux which is reflected back into space, the majority of solar radiation is absorbed by the atmosphere and the ocean. However, only 15% of the solar radiation is directly absorbed by the atmosphere and the remaining energy is converted into latent heat of condensation and fusion, turbulent heat conduction and long-wave radiation at the surface, to be subsequently released or absorbed into the atmosphere. From the first law of thermodynamics, the energy absorbed can reside as either internal (heat) energy or it can be used to do work against the environment, appearing as potential or kinetic energy. Consider a volume of fluid which is heated uniformly. The temperature, and therefore the internal energy, of the fluid will increase. At the same time the fluid will expand and do work against gravity. The potential energy of the fluid will therefore increase. The internal energy per unit mass (IE) is $C_v T$, where C_v is the specific heat at constant volume and T is the absolute temperature. The potential energy per unit mass (PE) is gz, where g is the acceleration due to gravity and z is the height above sea level. For a column of fluid of unit cross-section the mass is $\varrho\, \mathrm{d}z$, where ϱ is the fluid density and therefore:

$$\mathrm{IE} = \int_0^h C_v T \varrho\, \mathrm{d}z \qquad (9.1)$$

and

$$\mathrm{PE} = \int_0^h g z \varrho\, \mathrm{d}z \qquad (9.2)$$

where h is the depth of the fluid.

From the hydrostatic equation, $p = \rho gz$ and for the atmosphere, $p = \rho RT$ and thus:

$$PE = \int_0^h RT\rho \, dz \qquad (9.3)$$

The ratio of the potential energy to the internal energy for the atmosphere is therefore R/C_v or $\frac{2}{5}$. Thus for each unit of heat absorbed in the atmosphere $\frac{5}{7}$ of the heat will go into internal energy and $\frac{2}{7}$ will go into potential energy. If it is assumed that 70% of the incident solar radiation is absorbed both directly and indirectly by the atmosphere, then the fraction of solar radiation going into potential energy is $0 \cdot 7 \times \frac{2}{7}$ or $0 \cdot 2$ of the incident solar flux.

The ocean has a small compressibility and therefore virtually all of the absorbed solar radiation will appear as internal energy and only $0 \cdot 01\%$, a negligible fraction, will go into potential energy. This potential energy can be measured by sea level variations. For example, the seasonal cycle of temperature can produce changes in the sea level of a few centimetres.

Returning once more to the atmosphere, it is noted that, for the wind circulation to be in a steady-state balance, the energy dissipated by the wind must be balanced by an equal energy input which must, in turn, be derived from the potential energy of the atmosphere. The conversion of PE into kinetic energy (KE) is about $2 \, W \, m^{-2}$, from table 9.1, and this may be compared with a rate of $70 \, W \, m^{-2}$ for the production of PE. Therefore only 3% of the potential energy is used to drive the general circulation of the atmosphere. For the potential energy and internal energy to be in a steady-state balance, the majority of the potential and internal energy must be continuously and directly dissipated by long-wave radiation into space without being involved in the generation of kinetic energy. In thermodynamic terms, the atmosphere heat engine has a very low efficiency.

To understand why the atmosphere is so wasteful of its potential energy, consider two immiscible fluids of different density, ρ_1 and ρ_2, as shown in figure 9.1(a). The fluids are separated by a vertical partition and have equal volume. It can be shown that the potential energy of the system is $(gh^2/2)(\rho_1 + \rho_2)$, where h is the depth of the fluid. If the partition is now removed the denser fluid will flow under the lighter fluid and it will eventually take up the position shown in figure 9.1(b). During this process kinetic energy will be generated and it will, in its turn, be dissipated by the internal viscosities of the two fluids. In its final state the system still has potential energy but, because the interface is parallel to the geopotential surface, it is not possible to derive any more kinetic energy from the potential energy field. This remaining potential energy is known as unavailable potential energy and it can be shown to be equal to

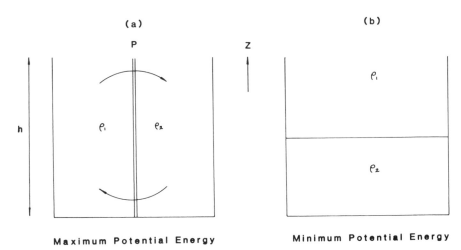

Figure 9.1. Potential energy in a two-layer fluid. (a) Maximum potential energy, before partition removed. (b) Minimum potential energy.

$(gh^2/4)(\varrho_2 + 3\varrho_1)$. The difference between the initial and final potential energy is $(gh^2/4)(\varrho_2 - \varrho_1)$. It is this fraction of potential energy which is converted into kinetic energy and it is known as the available potential energy (APE). The ratio of kinetic energy to potential energy is $\frac{1}{2}[(\varrho_2 - \varrho_1)/(\varrho_2 + \varrho_1)]$. If the density difference between the two fluids is small then the APE will be much less than the initial potential energy. In the atmosphere the horizontal variations in air density are relatively small when compared with the mean density, being typically less than 5%. Therefore the ratio of kinetic energy to potential energy is approximately 1%. In the ocean, where the above model is a better analogue, density variations are less than 0·5% and therefore the ratio of kinetic energy to potential energy is about 0·1%.

The APE is therefore proportional to the horizontal variations in density in both the atmosphere and the ocean. In the atmosphere the most satisfactory indicator of APE is the potential temperature because it avoids the adiabatic effect (figure 9.2). The horizontal gradient of potential temperature, and therefore APE, is caused by the difference in net radiation between the equator and the poles. The largest slope is in the middle latitudes. Most of the equatorial heating occurs at low levels whilst the polar cooling occurs at high levels in the troposphere. The potential temperature gradient is therefore between the lower equatorial troposphere and the upper polar troposphere. In the ocean the appropriate indicator of APE is the potential density although the potential temperature generally shows the main features of the distributions. The majority of the APE is associated with the slope of the main thermocline

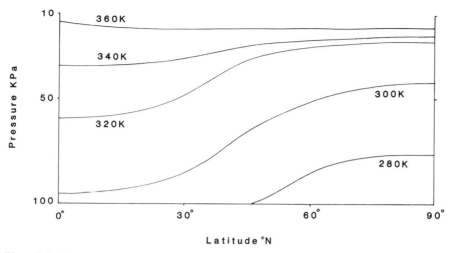

Figure 9.2. Longitudinally averaged atmospheric potential temperature.

which slopes upwards both polewards and equatorwards from the centres of the sub-tropical gyres. In the Atlantic and Pacific Oceans the largest APE is concentrated on the poleward boundaries of the gyres where the main thermocline rises from its deepest point to the surface in a horizontal distance of 200–300 km. Large slopes in the density surface also occur across the sub-tropical and Antarctic convergence zones, in the Antarctic Circumpolar Current. In deep water the density surfaces are close to horizontal and therefore, in common with the above analogue, there is little APE in the abyssal layers.

Figure 9.3 shows the energy cycle of the atmosphere and ocean in terms of the KE and APE reservoirs. For the atmosphere the APE is between three and four times larger than the KE and the total energy of the system is approximately $700 \times 10^4 \, \mathrm{J \, m^{-2}}$. If the energy of the atmosphere was not replenished by radiation, then all of its energy would be dissipated on a time scale of $(700 \times 10^4)/2 \cdot 3 \, \mathrm{s} \approx 35$ days which is very similar to the radiative time scale discussed in Chapter 2.

The APE and KE reservoirs of the ocean are not as certain as those given for the atmosphere and therefore figure 9.3 shows only some probable estimates obtained from a variety of investigations. The KE estimate is based on typical velocities of $10 \, \mathrm{cm \, s^{-1}}$ for the interior ocean and the APE is based on a model of the world ocean. It is noted that the ratio of APE to KE is approximately $25 : 1$ and therefore most of the energy of the ocean is stored as APE. This APE may be regarded as the flywheel of the ocean. However, the total energy reservoir of the ocean is less than 10% of the energy of the atmosphere and so the atmosphere is the energetically dominant system. There are two sources of energy for

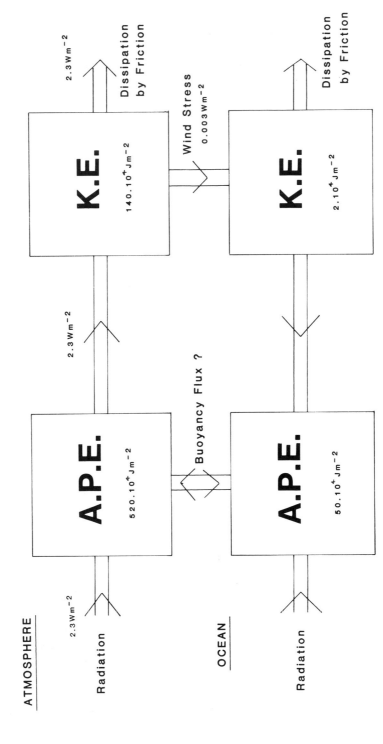

Figure 9.3. Energy diagram of the atmosphere and ocean. Note that the estimates of energy in the ocean are uncertain.

the ocean. One source is the direct driving of ocean currents by the wind stress. The other source is through the surface buoyancy flux which, in part, drives the thermohaline circulation. A surface wind of 10 m s^{-1}, typical of middle latitudes, will cause an energy transfer of approximately 1 W m^{-2} into the oceans. Of this energy input, some will be dissipated quickly in the surface layer, a fraction will go into surface waves and swell, some will go into internal waves and the remainder will go into ocean currents. The energy transfer into ocean currents is probably about 0·3% of the work done by the wind stress. The energy input by the surface buoyancy flux is not certain. One study has shown that the APE generated by surface cooling in the Northern Hemisphere is less than the energy input by the surface winds. Therefore the ocean is rather different from the atmosphere because the majority of its APE is derived from the kinetic energy input by the winds. Assuming that the dissipation rate is 0·3 × 10^{-2} W m^{-2} then the ocean would lose all of its APE and KE in about 5 years if it were not being continually replenished by the surface wind stress and surface buoyancy flux.

9.2. The kinetic energy of the atmosphere and ocean

The kinetic energy can be partitioned into two components:

(i) The mean kinetic energy (MKE), which is associated with the long-term average circulation of the ocean or atmosphere, thus:

$$\text{MKE per unit volume} = \tfrac{1}{2}\varrho\bar{v}^2$$

where $\bar{v}$ is the mean velocity.

(ii) The fluctuating or eddy kinetic energy (EKE), which is associated with the time variability of the currents and the winds, thus:

$$\text{EKE per unit volume} = \tfrac{1}{2}\varrho\overline{v'^2}$$

where v' is the fluctuation velocity.

The actual velocity at any given time, $v(t)$, is therefore given by $v(t) = \bar{v} + v'$.

In the atmosphere the MKE is approximately the same magnitude as the EKE whilst in the ocean the EKE is about 5–20 times larger than the MKE. For the observed kinetic energy in the atmosphere the typical mean velocity and fluctuation velocity would be 12 m s^{-1}. In the ocean the mean velocity is between 2 and 4 cm s^{-1} and the corresponding eddy velocity is 10 cm s^{-1}.

In both the atmosphere and the ocean there is a high correlation between the magnitude of the MKE and the EKE. This indicates that energy is being continually exchanged between the two modes. In the atmosphere the majority of the KE is associated with the locations of the jet streams in the upper troposphere between 30 and 50° latitude. The largest MKE occurs in the sub-tropical jet streams whilst the transient polar jet streams contribute most to the EKE. The equivalent high energy regions for the ocean are the western boundary currents and the equatorial currents but, in all of these regions, the fluctuating energy is four to five times larger than the MKE. This shows that even the most well-defined 'climatological' currents, such as the Gulf Stream, are regions of considerable variability. In the interior ocean, away from the boundary currents, most of the kinetic energy is in eddies and only a small fraction, usually less than 10%, is found in the mean flow. Gener-

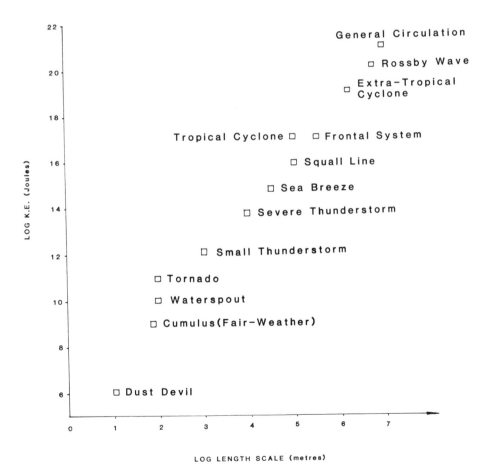

Figure 9.4. Kinetic energy of atmospheric phenomena.

ally both MKE and EKE decrease from the western to the eastern basins of both the Pacific and Atlantic oceans. In the Antarctic Circumpolar Current eddies are found along the entire length of the current and they are most intense near to the water-mass boundaries of the sub-tropical and Antarctic convergence zones.

It is interesting to compare the relative contributions of the different scales of atmospheric and ocean circulation systems to the overall budget of kinetic energy. Figure 9.4 shows the kinetic energy of a variety of atmospheric phenomena and figure 9.5 is a similar diagram for the ocean. It can be seen that the most intense atmospheric systems, such as the tornado and the tropical cyclone, are not the most energetic ones. An extra-tropical cyclone has a kinetic energy which is between 10 and 100 times larger than that of a tropical cyclone. The former system has a diameter about ten times as great as a tropical cyclone and so its larger kinetic energy is readily appreciated. As a further example, a sea-breeze system, which may penetrate inland for up to 100 km, has a kinetic

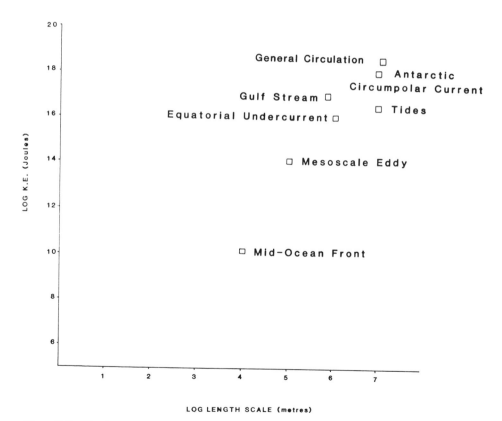

Figure 9.5. Kinetic energy of oceanic phenomena.

energy equivalent to about 10,000 tornadoes. Thus most of the fluctuation energy in the atmosphere is bound up in jet streams, Rossby waves and extratropical cyclones whilst the other scales of motion, although of local importance, make little contribution to the total kinetic energy budget. In the ocean the scale dependence of the circulation systems' contributions is again apparent. The Antarctic Circumpolar Current, the western boundary currents, such as the Gulf Stream and the Kuroshio, the equatorial undercurrents and the tides are the principal contributors whilst an individual mesoscale eddy, though locally energetic, has an energy which is only about 5×10^{-5} of the total kinetic energy. Thus 20,000 mesoscale eddies would be required in the ocean at any one time if their contribution was to be equivalent to the large-scale systems. Ocean eddies have a typical scale of 100 km and so the requirement would imply that virtually the whole ocean would have to be filled with eddies. As yet, it is not certain if this is the case but mesoscale circulations have been found in all of the ocean basins. Less energetic motions are associated with mid-ocean fronts and deep convection which occur on scales between 1 and 10 km. Surface waves and internal wave energy cannot be ignored in the kinetic energy budget. A wind of $10 \mathrm{~m~s}^{-1}$ will generate an equivalent surface-wave energy of approximately $2 \times 10^4 \mathrm{~J~m}^{-2}$. If this energy was typical of the whole ocean, it would be of a similar magnitude to the total kinetic energy of the circulation. It is also noted that internal waves have been shown to have a possible maximum energy of $\sim 10^3 \mathrm{~J~m}^{-2}$ and therefore they may also make a significant contribution to the total energy of the ocean circulation.

9.3. Mechanisms of kinetic-energy transfer

The simplest method by which kinetic energy is generated from available potential energy is by the mechanism in the tank experiment described in Section 9.1. In this experiment the dense fluid runs under the lighter fluid when the barrier is removed thus releasing kinetic energy. The vertical circulation that is set up in the tank is known as *direct circulation* because it always converts APE into KE. An example of direct circulation is the Hadley cell, shown in figure 6.7, where potentially warm air, driven by latent heat release, ascends into the ITCZ. The upward branch moves air polewards in the upper troposphere where it cools by long-wave radiation emission into space and then it decends in the sub-tropical latitudes. The buoyancy source in the lower equatorial troposphere, i.e. the release of latent heat by condensing water vapour, maintains the circulation and thus APE is converted into KE. A similar direct circulation can be visualized in the ocean where the sinking of dense water formed as the result of cooling or evaporation occurs, as in the Mediterranean Sea and the Red Sea (figure 5.9). Therefore direct circulations are associated with the upward motion of less-dense fluid and the downward motion of denser fluid.

Vertical circulations in which the denser fluid rises and the lighter fluid sinks are known as *indirect circulations*. In such cases, the system increases the APE at the expense of the KE. The Ferrel cell, in the middle latitudes of the troposphere (figure 6.7), is an example of an indirect, meridional circulation and acts as a sink to the KE of the atmosphere. The loss of KE due to the Ferrel cell is, however, small when compared with the KE gained from the Hadley circulation.

Indirect circulations are responsible for the generation of the large reservoir of APE in the ocean. In Section 7.5 it was shown that Ekman currents, produced by the large-scale wind circulation, cause surface convergence in the sub-tropical zones and divergence on the equator and in higher latitudes. The vertical Ekman circulation pushes warmer water downwards in the sub-tropical gyre and brings cold water up to the surface in the equatorial zone. It thus increases the APE. As has been shown, a surface-wind stress along the equator will produce a longitudinal tilt in the thermocline in both the Atlantic (figure 7.18(b)) and Pacific Oceans and this will also produce APE. In equatorial regions, because of the direct response of the thermocline to the wind, the APE reservoir can be replenished relatively quickly within a few months. In higher latitudes the process is rather less efficient and it would take in the region of a couple of decades to produce the observed tilt of the permanent thermocline.

The direct circulation is the simplest mechanism by which kinetic energy can be generated from available potential energy. In the atmosphere or in the ocean there is an additional constraint to the release of PE and that is the rotation of the Earth. Consider what would happen if the tank in figure 9.1 was rotating when the partition separating the two different density fluids was released. Initially the behaviour would be similar to the non-rotating case but, after a short time, the coriolis force would deflect the parcels of fluid perpendicular to the density gradient, as shown in figure 9.6. In the Northern Hemisphere the dense fluid would be deflected in a westward direction and the lighter fluid would be deflected in an eastward direction. Provided that the tank was an annulus, so that the motion was unimpeded perpendicular to the density gradient, then an equilibrium between the coriolis force and the pressure gradient would eventually be achieved. The zonal flows would be geostropic and the interface would slope as shown in the diagram. The rotation of the Earth therefore inhibits the conversion of APE into KE. The slope of the interface can be determined from the geostrophic and hydrostatic equations in the following way. If the geostrophic flow of the dense fluid is u_2 and that

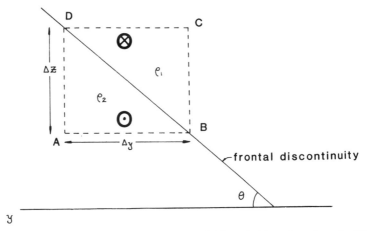

Figure 9.6. Geostrophic equilibrium for a two-layer fluid. y is directed northwards, ⊕ denotes eastward and ⊙ denotes westward geostrophic winds. Note rotation inhibits the adjustment of the fluid to a state of minimum potential energy. (See figure 9.1.)

of the lighter fluid is u_1, then:

$$f\varrho_1 u_1 = \frac{p_C - p_D}{\Delta y} \qquad (9.4a)$$

and:

$$f\varrho_2 u_2 = \frac{p_B - p_A}{\Delta y} \qquad (9.4b)$$

using the notation of figure 9.6. From the hydrostatic equation

$$p_B = p_C + g\varrho_1 \Delta z \qquad (9.5a)$$

and

$$p_A = p_D + g\varrho_2 \Delta z \qquad (9.5b)$$

Subtracting 9.5*b* from 9.5*a* and dividing through by Δy:

$$\frac{p_B - p_A}{\Delta y} = \frac{p_C - p_D}{\Delta y} - g(\varrho_2 - \varrho_1)\frac{\Delta z}{\Delta y} \qquad (9.6)$$

Substituting for the geostrophic equation (9.4*a* and *b*) into 9.6 to obtain the slope of the interface:

$$\frac{\Delta z}{\Delta y} = \frac{f(\varrho_1 u_1 - \varrho_2 u_2)}{g(\varrho_2 - \varrho_1)} \qquad (9.7)$$

As density variations are relatively small compared with velocity variations:

$$\varrho_1 u_1 - \varrho_2 u_2 \simeq \bar{\varrho}(u_1 - u_2)$$

where $\bar{\varrho} = (\varrho_1 + \varrho_2)/2$ and hence:

$$\frac{\Delta z}{\Delta y} = \frac{\bar{\varrho} f(u_1 - u_2)}{g(\varrho_2 - \varrho_1)} \qquad (9.8)$$

It can be seen that the interface, or front, will have a small slope when the density difference $(\varrho_2 - \varrho_1)$ is large and the velocity difference $(u_1 - u_2)$ is small. The interface will have a steep slope if the converse is true.

Consider the frontal interface between the northern boundary of the Gulf Stream, at approximately 20°C, and the Labrador Sea water at about 5°C. Assuming that the Gulf Stream water has an eastward velocity of $0 \cdot 5 \mathrm{~m\,s}^{-1}$ and that the cold water is quiescent then, from the equation of state:

$$\varrho_1 = 1024 \mathrm{~kg\,m}^{-3}$$
$$\varrho_2 = 1021 \mathrm{~kg\,m}^{-3}$$

assuming that the temperature dominates the density field. Hence from equation 9.8:

$$\frac{\Delta z}{\Delta y} = \frac{1022 \cdot 5 \times 10^{-4} \times 0 \cdot 5}{9 \cdot 8 \times 3} \simeq 1 \cdot 7 \times 10^{-3}$$

Thus the frontal surface will slope upwards towards the north by 170 m in 100 km.

For the atmosphere equation 9.8 has to be modified by substitution of the ideal gas equation: $p_1/RT_1 = \varrho_1$ and $p_2/RT_2 = \varrho_2$ and with the assumption that at the interface $\varrho_1 = \varrho_2$, the following expression is obtained:

$$\frac{\Delta z}{\Delta y} = \frac{f\bar{T}}{g}\left(\frac{u_1 - u_2}{T_2 - T_1}\right) \qquad (9.9)$$

where $\bar{T} = (T_1 + T_2)/2$. For the polar front $u_1 - u_2 = 30 \mathrm{~m\,s}^{-1}$, $T_2 - T_1 \sim 15$ K, $\bar{T} = 280$ K and

$$\frac{\Delta z}{\Delta y} = \frac{10^{-4} \times 280 \times 30}{9 \cdot 8 \times 15} \simeq 5 \cdot 7 \times 10^{-3} \quad \text{or} \quad \frac{1}{175}$$

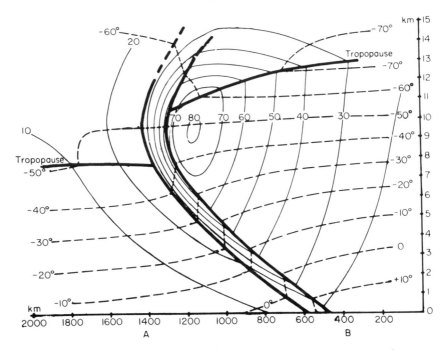

Figure 9.7. Schematic isotherms (dashed lines, °C) and isotachs (thin solid lines, m s⁻¹) in the polar front zone. Heavy lines are tropopauses and boundaries of frontal layer. (Reproduced from Atmospheric Circulation Systems, by E. Palmen and C. W. Newton, p. 176, figure 7.4, with permission of Academic Press.)

Therefore the frontal surface will have a calculated slope of $5 \cdot 7$ km in a horizontal distance of 1000 km which is similar to the observed value (figure 9.7).

It has been shown that the effect of the Earth's rotation is to produce regions of discontinuity of density in the ocean and of temperature in the atmosphere. These frontal regions also demarcate the boundaries of different air masses and different water masses. However, these frontal regions are rarely in equilibrium, as described by equations 9.8 and 9.9. Such a balance of forces would not allow for the transfer of heat and other properties across the frontal surface. As an extreme example, if the atmospheric transfer of heat across the $50°$ latitude circle was stopped then the atmosphere polewards of $50°N$ would cool by 100 K in 100 days. This increase in the horizontal gradient of temperature would in turn cause an increase in the upper-level winds relative to the low-level winds and, at a certain point, the wind shear between the upper and lower troposphere would reach a critical value and the flow would become hydrodynamically unstable. This is known as 'baroclinic instability' and figure 9.8(a) shows the initial deformation of a frontal surface as the result

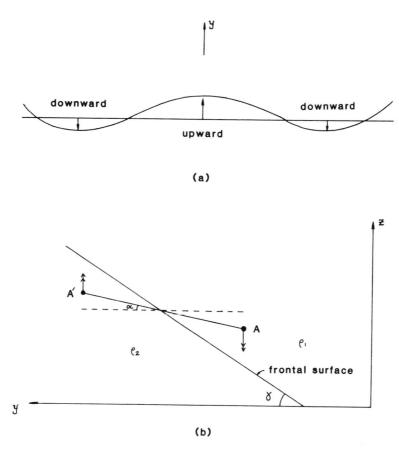

Figure 9.8. (a) Deformation of a frontal surface—plan view. (b) Unstable perturbations across a frontal discontinuity. Conversion of PE to KE will occur if $\alpha < \gamma$.

of this instability. The warm air moves polewards and upwards, over-riding the cold air, and the cold air moves equatorwards and downwards, undercutting the warm air. If the warm air follows the frontal surface, as shown in figure 9.8(b) an external supply of energy will be required to lift the warm air above the cold air. However, if the warm air follows trajectory AA' it will find itself in a colder environment and it will accelerate polewards and upwards without any energy input. Similarly a parcel of cold air following path $A'A$ will find itself in a warmer environment and it will accelerate equatorwards and downwards. This is very similar to the behaviour of air parcels in a vertically unstable environment where the air parcels continue to accelerate away from their initial positions (when once displaced). All of the parcels in the present example which have a trajectory between the horizontal and the frontal surface will gain kinetic energy from the APE of the front. This kinetic energy will produce a slantwise circulation across the isotherms. Figure 9.9 shows a more-detailed picture of the trajectories in an extra-tropical cyclone. The warm tongue can be seen to move upwards and polewards whilst polar

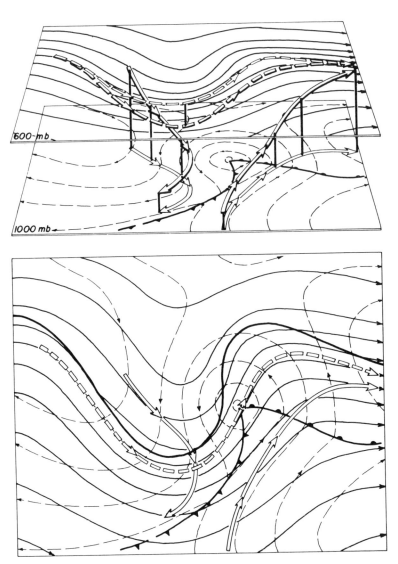

Figure 9.9. Perspective view of a 1000 mb cyclone and 600 mb contour pattern. Heavier arrows indicate three-dimensional trajectories in main ascending and descending branches; thin arrows, their projection onto a 1000 or 600 mb surface. (Reproduced from Atmospheric Circulation Systems, *by E. Palmen and C. W. Newton, p. 310, figure 10.20, with permission of Academic Press.)*

air, developing along the polar front in the rear of the system, can be seen to descend from high levels towards the surface thus releasing potential energy for the development of the extra-tropical cyclone. Latent-heat release in the region of ascent will provide an additional source of kinetic energy for the system. Eventually the tongue of warm air will be completely undercut by polar air and the system is said then to be occluded. At this stage most of the APE has been converted to kinetic energy and the extra-tropical cyclone will begin to decay as surface friction dissipates its kinetic energy. A typical extra-tropical cyclone will grow to its maximum intensity in 1–3 days and will decay in about 5–6 days. In some cases the development can be rapid—of the order of a few hours—and this makes accurate prediction of these systems very difficult.

Baroclinic instability is thus occurring at all times in the middle latitudes and it acts to transfer heat between low and high latitudes. However, this process does show a strong seasonal behaviour as well. In summer, with relatively weak horizontal gradients of temperature, the APE in the atmosphere is smaller and the extra-tropical cyclones are generally weaker. By contrast, in winter large horizontal temperature gradients are produced not only between the pole and the equator but also between the continents and the ocean. These large sources of APE produce intense cyclonic systems which transfer heat both polewards and towards the cold continental land masses. The major baroclinic zones in the Northern Hemisphere are located on the eastern seaboards of the North American and Asian continents in winter where very cold continental air occurs adjacent to the very warm Gulf Stream and Kuroshio Currents. In conclusion it is seen that vertical convection in low latitudes and slant convection in middle and high latitudes are the major processes by which energy is transferred in the atmosphere.

In the ocean, baroclinic instability is also responsible for the transfer of heat and other properties, such as salinity and nutrients, across frontal surfaces. Figure 9.10 shows the development of a Gulf Stream meander and the eventual formation of isolated eddies of cold and warm water on either side of the front. The growth rates of these meanders are slower than their atmospheric counterparts and they take between 1 week and 1 month to produce Gulf Stream rings. It is also noted that not all the meanders grown into eddies. The 'cold' Gulf Stream rings move southwards and westwards into the Sargasso Sea and they may be tracked for up to 2 years before they finally dissipate. In contrast, the 'warm' Gulf Stream rings, situated on the northern side of the Gulf Stream, tend either to be dissipated on the continental slope by friction or to be reabsorbed into the Gulf Stream system. Mesoscale ocean eddies are also important for the transfer of heat in the higher latitudes of the Southern Hemisphere. The Antarctic Circumpolar Current is the only zonal current which flows unimpeded around the globe and it therefore has some

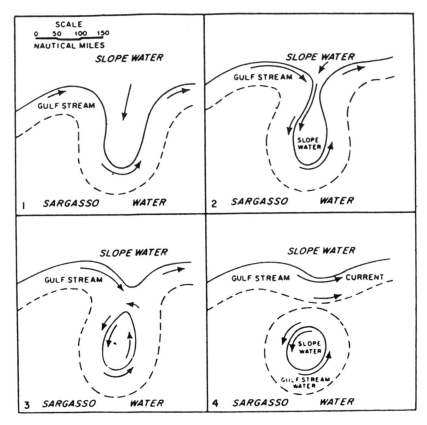

Figure 9.10. Diagram of Gulf Stream ring generation from meander formation to separation. (Reproduced from C. E. Parker (1971), Deep Sea Research, **18**(10), *982, figure 1, with permission of Pergamon Press.)*

similarities to the zonal atmospheric flows. Eddies are an important mechanism for transferring heat and salt across the zonal frontal boundaries. The energy for the eddies comes from the slope of the isopycnal (equi-density) surfaces between 40 and 60°S. Recent studies have shown that these Southern Hemisphere eddies may account for a horizontal heat transfer of $0 \cdot 4 \times 10^{15}$ W.

9.4. The general circulation of the atmosphere

The atmosphere and the ocean are both highly complex systems which fluctuate on a large variety of time and space scales. Both systems are dissipative and there is, therefore, a continuous flux of energy through the systems which maintains their circulations. Because of the dissipation of energy, the memories of the ocean and atmosphere are limited. In the atmosphere the KE can be dissipated in about 5 days if not replenished from the APE reservoir whilst in the ocean the KE can be dissipated within 100 days. These limited dynamical memories imply that the predictability of the behaviour of, say, an extra-tropical cyclone or an ocean eddy is limited by the dissipation time scale. It is therefore quite impossible to describe the behaviour of the atmosphere or ocean in terms of the influence of an individual eddy circulation. The only possible approach is to evaluate the statistical influence of these highly variable eddies on the behaviour of the general circulation.

To describe the behaviour of the general circulation of the atmosphere it is necessary to concentrate attention on two quantities:

(i) The energy of the atmosphere and, in particular, the transfer of energy between the equator and the pole.
(ii) The momentum balance of the atmosphere. This is important because it gives an insight into the reasons for the observed distribution of wind, as shown in figure 6.5, and also into why the sub-tropical jet streams are located at $30°$ latitude. It also yields information about the role of extra-tropical cyclones in maintaining the climatological wind distribution.

A poleward flux of energy is required by the ocean–atmosphere system to counteract the net radiation imbalance between the poles and the equator. The atmospheric energy, E, can be written as:

$$E = KE + IE + PE + LE \qquad (9.10)$$

where KE is the kinetic energy, IE is the internal energy, PE is the potential energy and LE is the latent heat of the atmospheric water vapour. The KE of the atmosphere is small compared with the other energy terms and it is generally ignored in the total energy budget. The IE and PE are usually grouped together as the potential heat flux.

The change in the energy of the atmosphere with time, dE/dt, is given by:

$$\frac{dE}{dt} = Q - D \qquad (9.11)$$

where Q is the energy source and D is the energy dissipation by frictional processes. The diabatic heating term, Q, is composed of the net radiation heating, the sensible heat flux from the surface and the latent heat released by condensation. For the atmosphere as a whole the dissipation, D, must balance the net heat input, Q. However, between the equator and the pole, the equation does not balance at every point and therefore

a transfer of energy is required to maintain the distribution of heat
sources and sinks. Figure 9.11 shows the meridional distribution of Q in
the Northern Hemisphere winter. Everywhere in the troposphere the
atmosphere is cooled by longwave radiation at a rate of between 1 and
2 K day^{-1}. Sensible heat flux from the surface heats the lower 1–3 km of
the troposphere, especially in the winter hemisphere, and latent-heat
release is responsible for heating all of the troposphere in the equatorial

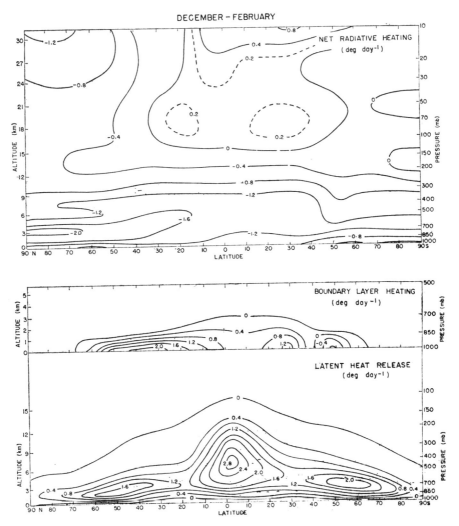

Figure 9.11. *Components of diabatic heating for the atmosphere for December–February. Units:*
°C day^{-1}. *(Reproduced from R. E. Newell* et al., *p. 63, figure 7, in* The Global Circulation
of the Atmosphere, *edited by G. A. Corby, with permission of the Royal Meteorological Society.)*

zone and for heating the lower and middle troposphere at higher latitudes. The net heating distribution shows the whole equatorial zone and the surface boundary layer as net heat sources whilst the remaining regions are net heat sinks. To maintain this distribution energy has to be transferred both polewards and upwards by the general circulation. The horizontal flux of energy, F_A, across a latitude circle is given by:

$$F_A = \frac{2\pi a \cos \phi}{g} \int_{p_0}^{p_s} [\overline{vE}] \, \mathrm{d}p \qquad (9.12)$$

where v is the northward velocity, E is the energy, $2\pi a \cos \phi$ is the distance along a latitude circle. The bar denotes a time average and the square brackets denote a zonal average.

The mean northward transport of energy can be partitioned into two terms:

(i) $\overline{v}\overline{E}$—the transport of mean energy by the average meridional circulation such as the Hadley and Ferrel cells.
(ii) $\overline{v^1 E^1}$—the transport by the fluctuating component of the meridional circulation, or the eddy flux. This term includes fluctuations caused by the day-to-day variations in the position of the Rossby waves in the upper troposphere and by synoptic-scale motions in the lower troposphere.

Figure 9.12 shows the contributions of the potential heat flux and the latent heat flux to the total heat transport. It can be seen that the mean meridional circulation is very effective at transferring heat at low latitudes but it is less efficient in middle and high latitudes. The Ferrel cell in the middle latitudes actually transfers heat towards the equator against the temperature gradient. At middle and higher latitudes all of the poleward heat flux is achieved by the eddy circulations. These eddy circulations have not only to transfer the heat demanded by the distribution of heat sources and sinks but they also have to transfer an extra amount to counterbalance the equatorward transport of heat by the mean circulation. The latent heat flux shows a different behaviour from that of the potential heat flux. At low latitudes the Hadley cell transports water vapour towards the equator where it is required because of the net deficit of water in the rising branch of the Hadley cell, as described in Section 5.8. At higher latitudes the eddy flux of water vapour, like the eddy heat flux, is much larger than the mean meridional circulation. However, because the majority of the water vapour is confined to the lowest 5 km, the Ferrel circulation actually produces a latent heat flux in the same direction as the eddy transport. In addition, the major contribution of the eddy latent heat flux is made by the synoptic-scale motions in the lower atmosphere. The total transport of energy by the atmosphere necessary

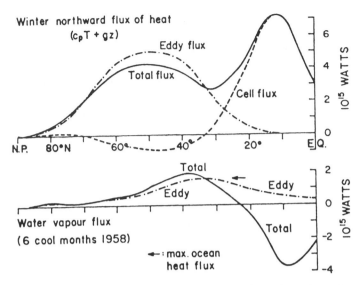

Figure 9.12. Upper diagram shows northward potential heat flux in northern winter. Lower diagram shows water vapour flux. Arrow shows latitude and magnitude of strongest northward heat flux by ocean currents (an annual average). (Reproduced by C. W. Newton, p. 138, figure 1, in The Global Circulation of the Atmosphere, *edited by G. A. Corby, with permission of the Royal Meteorological Society.)*

to balance the distribution of heat sources and sinks can be obtained by the summation of the potential heat flux and the latent heat flux.

In conclusion, it has been shown that the variability of the atmosphere is not purely random but that it acts in a systematic manner to transfer energy from low to high latitudes, and eddy circulations play an important role in this process. A similar approach is also necessary to explain the momentum balance of the atmosphere.

The total angular momentum of the Earth, including the atmosphere and the ocean, is constant. Any changes in the angular momentum of one component of the system must be balanced by a corresponding change in the angular momentum of another of the system's components. The atmosphere exchanges angular momentum with the solid Earth by the frictional torque that it exerts on the Earth's surface. If this net torque were in the direction of rotation then the Earth would increase its angular momentum, and hence its rotation rate, at the expense of the atmosphere's angular momentum. Correspondingly, a net torque in the opposite direction would slow the Earth's rotation rate and the atmosphere would gain angular momentum. Although small variations in the angular velocity of the Earth do occur it can be assumed that, over a period of 1–2 years, the angular velocity is constant. Hence the total angular momentum of the atmosphere should be in balance when it is averaged over several seasonal cycles.

The net torque on the Earth's surface does vary with latitude. At low latitudes the winds have a slower rotation rate than the Earth and therefore the surface torque acts in the opposite direction to the Earth's rotation. The Earth will lose angular momentum and the atmosphere will gain eastward angular momentum at these low latitudes. In the middle latitudes the surface winds have a higher rotation rate than the Earth and they produce a torque in the same direction as the Earth's rotation. In this case, the atmosphere will lose and the Earth will gain angular momentum. There will, therefore, be a sink of atmospheric angular momentum in the middle latitudes and a source in low latitudes. In order to maintain this distribution of sources and sinks the meridional component of the atmospheric circulation must transport angular momentum from low to middle latitudes. Over the polar regions the surface winds are westwards and these areas therefore act as a small source of angular momentum which will also be transferred to middle latitudes.

The angular momentum per unit mass of a parcel of air is given by

$$a \cos \phi u + a^2 \cos^2 \phi \Omega \qquad (9.13)$$

where u is the zonal component of the wind, Ω is the rotation rate of the Earth, a is the radius of the Earth and ϕ is the latitude.

The first term is the relative angular momentum of the zonal wind and the second term is the angular momentum of the Earth. Furthermore, the annual average of the northward flux of angular momentum, $M\phi$, integrated over the depth of the atmosphere, can be shown to be:

$$M\phi = \frac{2\pi a \cos \phi}{g} \int_0^{p_0} [\overline{uv}] a \cos \phi + \Omega [\overline{v}] a^2 \cos^2 \phi \, dp \qquad (9.14)$$

The overbar indicates a time average, the square brackets indicate a zonal average and v is the meridional component of the wind.

It will now be shown that the total transport of angular momentum consists only of the first term in equation 9.14. Consider a column of the atmosphere extending around a latitude circle. A northward mass transfer into the column at one level must be balanced by an equal southward mass transfer out of the column at a different level in order to avoid a net accumulation of mass in the column. Hence the total mass transport over the whole atmospheric column is zero and integration of the zonal average meridional velocity $[\overline{v}]$ over the column is also zero. Therefore:

$$M\phi = \frac{2\pi a^2 \cos^2 \phi}{g} \int_0^{p_0} [\overline{uv}] \, dp \qquad (9.15)$$

Now the term $[\overline{uv}]$ is the time-averaged correlation between the zonal wind and the meridional wind. It can be partitioned into two components in a similar manner to the potential heat and latent heat water vapour flux. The first component is the contribution of the mean circulation, $[u]\,[v]$, and the second is the contribution of the eddy circulation, $[\overline{u'v'}]$, where u' and v' are the departures from the mean wind. The northward transport of angular momentum from low to middle latitudes, shown in figure 9.13, which is required to maintain the surface torque balance, is dominated by the eddy flux, $[\overline{u'v'}]$, and it is only in equatorial latitudes where the Hadley cell contributes an equivalent mean transport. The Hadley cell produces a positive contribution to the $[\bar{u}]\,[\bar{v}]$ term, since the poleward branch of the cell (i.e. $[\bar{v}] > 0$) is associated with an eastward wind (i.e. $[\bar{u}] > 0$), whilst the low-level equatorward flow (i.e. $[\bar{v}] < 0$) has westward winds for which $[\bar{u}] < 0$. Therefore the product $[\bar{u}]\,[\bar{v}]$ is positive in both cases and the Hadley cell is responsible for a poleward transport of angular momentum.

The Ferrel cell produces a small equatorward transport of angular momentum and therefore, as with the potential heat flux, the eddy momentum flux is the only mechanism available to produce the required poleward momentum flux. The maximum eddy momentum flux occurs in the upper troposphere a few degrees equatorwards of the sub-tropical jet streams. It is this convergence of the eddy momentum flux which provides the source of angular momentum to maintain the jet streams.

Although the meridional cells are not the major contributors to the

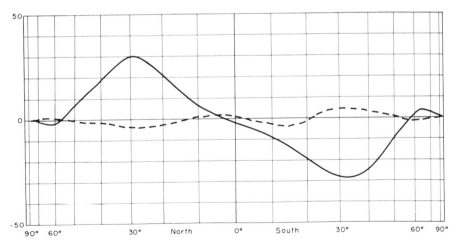

Figure 9.13. Northward transport of angular momentum by eddies (solid curve) and mean meridional circulation (dashed curve). (Reproduced from The Nature and Theory of the General Circulation of the Atmosphere, *by E. N. Lorenz, p. 82, figure 40, with permission of the World Meteorological Organization.)*

meridional flux of angular momentum, they are important in the *vertical exchange* of momentum. The Earth's angular momentum, represented by the 'Ω' term in equation 9.14, is a maximum at the equator and decreases towards the poles. The ascending branch of the Hadley cell has a larger angular momentum than the descending branch at a higher latitude. The Hadley cell will therefore produce a net upward flux of eastward momentum. The Ferrel cell acts in the opposite direction and brings eastward momentum down to the surface. Vertical eddy fluxes, associated with large convection cells in low latitudes and synoptic-scale systems in higher latitudes, will also contribute to the vertical flux of angular momentum.

It is seen, therefore, that both eddy fluxes and the mean meridional air circulations contribute to the horizontal and vertical transfer of angular momentum necessary to maintain the observed distribution of surface wind. The eastward winds at middle latitudes and westward trade winds at low latitudes are a necessary consequence of the conservation of angular momentum. Furthermore, because the torque is given by $a \cos \phi \tau_0$, where τ_0 is the surface wind stress and ϕ is the latitude, the eastward winds at middle latitudes must generally be stronger than the westward winds at low latitudes in order to maintain the torque balance.

Recent studies on the momentum balance of intense current systems in the *ocean*, such as that of the Gulf Stream, have also shown that eddy circulations are very effective at pumping angular momentum from the surface layers into the deep ocean. The addition of momentum to these deeper regions produces well-organized, quasi-permanent deep gyre circulations under the Gulf Stream which are larger than the individual mesoscale eddies. Therefore in both the ocean and the atmosphere the transient eddies not only contribute to the exchange of heat and other tracers but are also instrumental in maintaining the observed mean circulation.

9.5. Models of the atmosphere and the ocean

A meteorologist or a physical oceanographer is at a distinct disadvantage compared with scientists in many other disciplines for he has only one laboratory and that is the atmosphere or the ocean! Experimental programmes, carefully designed to study a particular scale of motion, such as a tornado or a mesoscale eddy, are themselves exposed to the vagaries of the atmosphere and ocean. Often the measurements obtained are incomplete and definitive conclusions cannot always be drawn. A well-designed current meter array may be rendered useless by the loss of moorings in a strong current or a storm. Despite improvements in the reliability of technology, the lack of control over experimental conditions is a unique factor of the science of meteorology and oceanography.

This basic experimental difficulty has spurred the use of laboratory models of both the atmosphere and the ocean. These models are of two types. The first type is a physical analogue model. For instance, an analogue of the atmospheric general circulation can be obtained by rotating an annulus containing a fluid on a turntable. The temperature gradient between the pole and the equator is simulated by heating the outer rim of the annulus and by cooling its inner rim. The experiment facilitates the study of the behaviour of the flow under controlled conditions. The experimental scientist is able to change the rotation rate and temperature gradient at will and is able to repeat the experiment many times to check the validity of the results.

The second type of model is the numerical analogue which uses the computer. These models are based on the numerical solution of a set of differential equations which approximate closely to the equations of motion on a rotating Earth (i.e. equations 7.17 and 7.18). These models can be used to study a variety of scales of motion, from the general circulation of the atmosphere and ocean to the detail of a thunderstorm or tornado. Again the experimenter has the advantage that the parameters in the model can be changed at will in order to gain a deeper insight into the mechanisms of the phenomena under study. Furthermore, the modeller has a complete set of calculated data from which the energy and momentum balance of the system can be deduced. Models of large and synoptic scales of the atmosphere are now used routinely for weather prediction for up to 10 days ahead. General circulation models, which include, in addition to the equations of motion, numerical analogues of physical processes such as radiation absorption and emission, clouds and surface processes, can be used to study the longer-term effects of sea surface temperature anomalies, volcanic aerosols and changes in the atmospheric constituents, such as carbon dioxide. In the last two decades the numerical model has indeed become a very powerful tool for understanding the behaviour of the atmosphere and ocean.

A simplified numerical model of the atmosphere will now be considered. It is of the type used for weather prediction and is presented in order to illustrate the basic procedures. The atmosphere is first gridded in three dimensions. There are usually between 5 and 15 levels in the vertical dimension and the majority of these levels are in the troposphere. In the horizontal dimension the resolution has to be between 100 and 200 km in order to resolve synoptic-scale circulations such as cyclones and anticyclones. The observed state of the atmosphere at time t is obtained from synoptic weather stations using radiosonde balloon measurements. This information, which includes the temperature, humidity, pressure and the horizontal components of the wind, is then interpolated onto the three-dimensional grid. For a 100 km grid covering the Northern Hemisphere with 10 levels, about $2 \cdot 5 \times 10^6$ numbers are required to describe the initial state of the atmosphere. The x component of the equation of motion can be written as:

$$\frac{\partial u}{\partial t} = F(u, v, w, p) \tag{9.16}$$

where $\partial u/\partial t$ is the local acceleration of velocity. The right-hand side includes the accelerations arising from the pressure gradient, the coriolis force and inertial forces. From the initial data all these accelerations can be obtained by numerical approximation. For example, the pressure-gradient acceleration, $(1/\varrho)(\partial p/\partial x)$ at a point i, can be approximated by $(1/\varrho_i)[(P_{i+1} - P_{i-1})/2\,\Delta x]$, where Δx is the grid spacing. The net acceleration, F, can thus be obtained at each grid point at time t. Furthermore, by numerical approximation, equation 9.16 can be written as:

$$\frac{u(t + \Delta t) - u(t)}{\Delta t} = F(t)$$

Hence:

$$u(t + \Delta t) = u(t) + \Delta t F(t) \tag{9.17}$$

As $F(t)$ and $u(t)$ are known, u can be calculated at a later time, $t + \Delta t$. The time step, Δt, would usually be between 10 and 30 min. A similar procedure can be applied to the y component of motion and to the other equations for the vertical motion, temperature, pressure and humidity to obtain values of all the variables at the later time $t + \Delta t$. By repeated application of equations similar to equation 9.17, values at $t + 2\,\Delta t$, $t + 3\,\Delta t$, etc., are thus obtained. For a time step of 10 min, 144 cycles would be necessary to obtain a forecast for one day ahead.

In principle, it would seem that the calculation could be extended indefinitely into the future to give forecasts of not only a few days ahead but for a month, or even a year, ahead. However, there are both practical and theoretical limitations to prediction. The first limitation lies in the errors in the prescription of the initial state of the atmosphere. These errors are associated with interpolation of data onto the grid, particularly in regions of sparse data such as over the ocean. The second limitation is that the grid can only resolve scales of motion which are larger than the grid spacing and therefore systems such as squall lines, thunderstorms and cumulus convection cannot be included in the prediction. The non-linearity of the atmosphere implies that these unresolved scales may transfer energy into the larger-scale systems and thus change the development of a synoptic system. Lastly, the atmosphere is dissipative and therefore it cannot retain information of the previous history of the atmospheric state for more than about 10–14 days into the future. Hence all numerical predictions will degrade as the integration proceeds. An analogous situation arises in the transfer of information where small errors introduced during repeated transcription can eventually produce a meaningless message.

General circulation models of the atmosphere can be integrated for much longer periods than the dissipative time scale, ranging from months to many years. Here, the interest is not in the prediction of an individual cyclone or anticylone but in the long-term statistics of the flow. The information may be time averaged to produce maps of surface pressure, velocity and temperature, similar to standard climatological maps shown in figures 6.5 and 6.6. In addition, the momentum balance and energy balance may also be analysed for comparison with observed estimates. This type of model is very much a research tool for understanding the behaviour of the atmosphere but it is anticipated that, eventually, these models will be used to make monthly and seasonal predictions. These predictions will inevitably be different from short-term weather forecasts because they will be statistical rather than deterministic in nature.

Similar methods to those used in atmospheric models have been adopted for ocean models. However, because the energetic scales of motion are in the range of 50–200 km, much higher resolution is required. Horizontal grids of 10 km are necessary to resolve an ocean eddy field. This means that a hundred times more calculations are required for a given area than for the equivalent atmospheric calculation. Furthermore, the time step, Δt, has to be reduced as the resolution is increased thus further increasing the number of computations. For this reason it is only relatively recently that eddy-resolving ocean models have been developed for realistic ocean geometries. Even so, most eddy models only consider the behaviour of a small part of the ocean or an idealized basin. In contrast, general circulation models of the ocean have been under development for a longer period of time. These models are either applied to flows in individual ocean basins or to the world ocean and they have to be integrated for simulation times of up to 1000 years to reach equilibrium. Although in one sense they are the counterparts of atmospheric general circulation models, their resolution of 100–200 km precludes the inclusion of the mid-latitude mesoscale eddies and the effects of the eddies have to be parametrized as a sub-grid scale process. Their main use is in the understanding of the distribution of water masses and tracers as well as for long-term climate experiments.

10

Climate variability and predictability

10.1. *Air−sea interaction: an introduction*

Air-mass transformation over the ocean

The modification of a continental air mass as it moves over the ocean is one of the more important processes necessary to the understanding of the interaction between the ocean and the atmosphere. In order to illustrate these processes two examples of air-mass transformation will be considered.

The first example is the transformation of the trade winds on their journey from the sub-tropical land masses to the ITCZ. Figure 10.1 shows a schematic diagram of some of the salient features of the trade-wind system between the Saharan coast and the equatorial zone. Dry, warm surface air from the continent is rapidly transformed by contact with the cooler surface water, which is at a temperature of approximately 15°C, in a shallow boundary layer. The moisture content in the boundary layer is rapidly increased by evaporation from the ocean whilst sensible heat is lost from the air to the ocean because of the reversal of the vertical temperature gradient. In addition, the air is cooled by long-wave radiation into space. The surface air therefore cools rapidly to the sea surface temperature and, because of the addition of water vapour and the net cooling, low stratus clouds form in the boundary layer. At this stage the boundary layer may be relatively shallow having a depth of about 500 m. Above the maritime boundary layer the air is potentially warmer and very dry because of the descent of air from high levels in the troposphere associated with subsidence in the tropical anticyclones. Therefore, at the boundary between the two air masses, a strong trade-wind inversion is formed which tends to suppress vertical motion and mixing. As the maritime air moves over progressively higher sea surface temperatures, the latent heat flux increases and the sensible heat flux becomes upwards from the ocean to the atmosphere. The warming of the air mass results in a break-up of the stratus cloud and the development of trade-wind

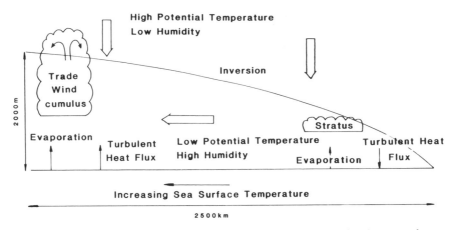

Figure 10.1. Schematic diagram of the modification of the trade wind boundary layer over the ocean.

cumulus clouds. The shallow cumulus convection mixes the moist surface air with the dry air above the inversion. This results in a deeper boundary layer. The convection also brings drier air towards the surface which, in turn, increases the evaporation from the ocean into the boundary layer. In the western ocean basins the boundary layer reaches a depth of 2–3 km and evaporation is three times larger than in the eastern ocean basins. In these boundary layers convection is deep enough to produce showers and this warms the cloud layer by latent heat release. Energy budgets of the trade-wind boundary layer have shown that the downward heat flux through the inversion is as large as the upward surface flux of sensible heat. They have also shown that virtually all of the water evaporated into the boundary layer is exported horizontally out of the trade-wind zone and is subsequently released in the tropical convergence zones. A typical trade-wind velocity is 6 m s^{-1} and therefore the surface air mass, providing that it remains in the boundary layer, will traverse a distance of 2500 km between the source region and the equatorial region in about 5 days.

The second example of air-mass transformation is more typical of air–sea interaction in middle latitudes. During winter very cold, dry air from the northern Asian and North American land masses periodically surges out over the relatively warm waters of the western North Pacific and North Atlantic Oceans. These cold outbreaks occur in the north-westerly flow in the rear of an extra-tropical cyclone and they may last for several days. Figure 10.2 shows the transformation of a cold, dry stable air mass as it flows out over the East China Sea and the warm Kuroshio current. Initially the continental air mass may have a temperature between 0 and $-20°$C and a very low specific humidity of

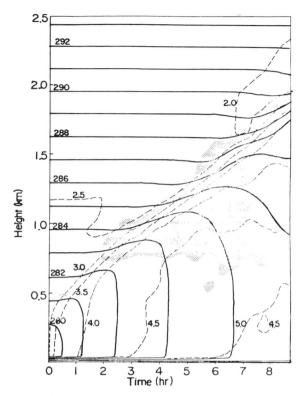

Figure 10.2. Simulation of the time evolution of a moist convectively mixed layer in a cold outbreak over the Kuroshio current. Solid line indicates potential temperature, K, dashed line specific humidity, g kg⁻¹ and stippled area cloud layer. (Reproduced from WMO, Garp Publication Series, 24, p. 167, figure 6.1, with permission of WMO.)

less than $1 \, \mathrm{g \, kg^{-1}}$. The sea surface temperature, in contrast, increases from about $5 \,^{\circ}\mathrm{C}$ near the coast to $20 \,^{\circ}\mathrm{C}$ over the Kuroshio current. As the air mass first moves out across the East China Sea, the unstable boundary layer will be rapidly warmed by an upward flux of sensible heat. Although the air is dry, it is initially very cold and is therefore unable to absorb much moisture. As the boundary layer warms and deepens the surface latent heat flux becomes comparable with the sensible heat flux. In contrast to the trade-wind boundary layer, which deepens only slowly, these cold air outbreaks cause rapid changes in the depth of the boundary layer. Over a spatial scale of 300 km depth changes of 1–2 km are observed. During the Air-Mass Transformation Experiment (AMTEX) in 1974 and 1975, total surface heat fluxes of $700-800 \, \mathrm{W \, m^{-2}}$ were observed over two separate 4 day periods. This heat flux, which is half the solar constant (i.e. $1360 \, \mathrm{W \, m^{-2}}$), would heat a column of air 2 km deep at a rate of $30 \, \mathrm{K \, day^{-1}}$ and would lead to vertical instability of the boundary layer.

The growth of convective cloud is limited by the height of the inversion which marks the boundary between the modified air mass below and the continental air mass above. Near the coast the boundary layer depth is small and therefore the convective cloud is shallow being usually stratocumulus, and of small horizontal dimensions. However, as the boundary layer grows, both the depth and horizontal scale of the convective cloud increases and it can usually grow sufficiently to produce

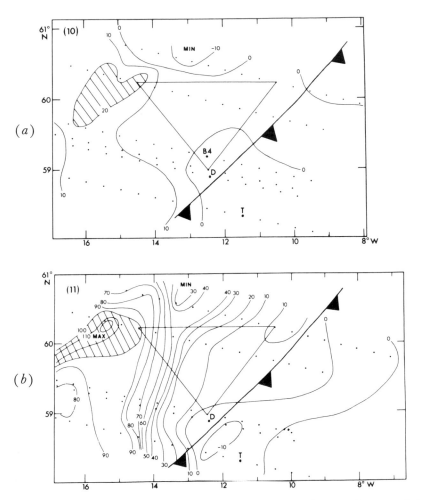

Figure 10.3. (a) Horizontal variation of the surface turbulent heat flux over the ocean, during the passage of a cold front, (Q_H: $W\,m^{-2}$), at 01.00 hours GMT, 31 August 1978. Values of more than 20 $W\,m^{-2}$ are shown hatched. (b) Horizontal variation of the surface latent heat flux (Q_E: $W\,m^{-2}$) at 01.00 hours GMT, 31 August 1978. Values of more than 100 $W\,m^{-2}$ are shown hatched. (Reproduced from Guymer et al. (1983), Phil. Trans. Roy. Soc., **A308**, 265, figures 10 and 11, with permission of the Royal Society.)

showers. The regions of large air-mass transformation are also regions of frequent cyclonic development, as the result of the larger gradient of surface temperature. The region south of Japan is the most active region of cyclone generation in the world during the winter season. The average heat loss of the ocean over the Kuroshio current in February is 350 W m^{-2} which is sufficient to cool the upper 100 m layer of the ocean by $2 \cdot 2°C$ in 1 month.

The two previous examples have shown that the extraction of heat from the ocean is strongly dependent on the temperature and humidity characteristics of the overlying air mass relative to the sea surface temperature. In an extra-tropical cyclone two or three different air masses may cross a given area in a day and produce large variations in surface heat fluxes. Figure 10.3 shows the surface fluxes of latent and sensible heat during a period of relatively weak air–sea interaction in summer in the north-eastern Atlantic Ocean. The cold front marks the boundary between a cooler, drier, air mass to the north-west and a warm air mass to the south-east. In the warm sector both latent heat and sensible heat fluxes are close to zero and the air mass is therefore receiving little or no heat from the ocean. This is typical for warm, humid air masses moving over progressively lower sea surface temperatures. In the rear of the cold front the latent heat flux increases to a maximum of 110 W m^{-2} about 300 km from the front whilst the sensible heat flux is much less, being in the range from 10 to 20 W m^{-2}. This contrasts strongly with the cold out-breaks over the Kuroshio current where sensible heat fluxes are of a similar magnitude to latent heat fluxes. It will also be noticed that, in the rear of the cold front, there are variations in latent heat flux caused by horizontal variations in the sea surface temperature of $1°C$.

Response of atmospheric phenomena to sea surface temperature

All scales of atmospheric motion are involved in the transfer of energy between the ocean and atmosphere and therefore most atmospheric phenomena will respond, directly or indirectly, to the sea surface temperature. On the small scale an increase in sea surface temperature may change the stability of the overlying air mass which may, in turn, change the boundary layer cloud from a stable stratus type to an unstable convective type. Similarly, the distribution of sea fog may be affected by a relatively small change in the sea surface temperature. Mesoscale circulations may also be influenced by the sea surface temperature. The intensity of a sea-breeze circulation depends on the difference in temperature between land and sea to generate the available potential energy which sustains the circulation. Mesoscale convection over the ocean will depend on the air–sea temperature difference to produce latent and sensible heat fluxes to sustain the vertical circulation. In turn, the vertical circulation will bring colder and drier air to the boundary layer and this will enhance the vertical transfer of heat into the circulation.

The tropical cyclone is an example of the self-sustaining circulation described above. The generation areas of the tropical cyclone are usually limited to the regions of the tropics where sea surface temperatures are above $26\,^{\circ}$C. Evaporation rates are determined principally by the surface wind and by the sea surface temperature. It is also noted that the vapour pressure has a non-linear dependence (figure 4.9) and thus high evaporation rates occur over the tropical ocean. An embryonic low-pressure system will produce a convergence of water vapour in the maritime boundary layer which will release latent heat when uplifted. As the horizontal and vertical circulation gains kinetic energy from the latent-heat release, the surface evaporation rate and the surface convergence of water vapour will both increase and the circulation will intensify. Provided that there is a copious supply of water vapour, the storm will increase in intensity at an exponential rate until the loss of kinetic energy by surface friction balances the kinetic energy input. A mature tropical cyclone can produce surface latent heat fluxes of $1000\,\mathrm{W\,m^{-2}}$ and sensible heat fluxes of $500\,\mathrm{W\,m^{-2}}$. These are approximately twice the surface heat flux observed in cold outbreaks over the Kuroshio current. The tropical cyclone is a truly oceanic phenomenon because, when it strikes land, its surface moisture source is reduced and this, in combination with higher surface friction, will cause it to dissipate quickly. Occasionally tropical cyclones will propagate into higher latitudes where they may trigger the development of an extra-tropical cyclone. Over the higher latitude oceans the extra-tropical cyclone may increase its energy by ventilating the marine boundary layer with drier air which has subsided from the upper troposphere in the rear of the cold front. The dry air is able to pick up heat and moisture which can both be recirculated into the ascending region of the cyclone. They may be released subsequently as available potential energy and kinetic energy.

The larger-scale circulation of the atmosphere is strongly influenced by the distributions of the continents and oceans. In middle latitudes the climatological position of the long Rossby waves is determined, in part, by the distribution of heat sources and sinks. Figure 10.4 shows the observed surface pressure and 500 mb height as well as surface heat sources and sinks at middle latitudes of the Northern Hemisphere. The principal feature predicted by theory (Smagorinsky, 1953) is the development of a surface anticyclone (surface cyclone) about $25\,^{\circ}$ downwind of the heat sink (heat source) and a ridge (a trough) at 500 mb displaced by $10\,^{\circ}$ to the west. During winter the North Atlantic and North Pacific Oceans provide an average surface heat source of $100\,\mathrm{W\,m^{-2}}$. The largest heating is located over the Kuroshio and Gulf Stream currents, as seen in the previous subsection. In contrast, the continental land masses are net heat sinks in winter

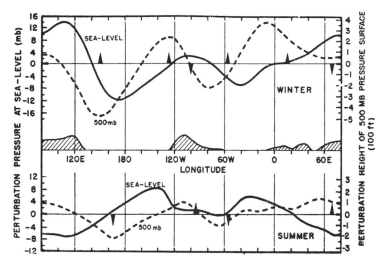

Figure 10.4. Zonal profiles of normal perturbation pressure at sea level (solid line) and normal perturbation height of the 500 mb pressure surface (dashed line). Continental elevations are indicated schematically by a cross-hatched area. Positions of observed relative heat sources and sinks indicated by spires pointed upwards and downwards, respectively. Upper part: winter profiles averaged over 20° latitude centred at 45°N. Lower part: summer profiles averaged over 20° latitude centred at 50°N. (Reproduced from J. Smagorinsky (1953), Q. J. Roy. Met. Soc., 79(341), 362, figure 7, with permission of the Royal Meteorological Society.)

because heat loss by long-wave radiation exceeds the input by solar radiation. According to theory surface low-pressure centres should be expected about 25° downstream of the heat sources in the north-western Atlantic and north-western Pacific. This is in rough agreement with the observed positions of the Icelandic and Aleutian lows. Over the Eurasian continent the Siberian anticyclone should be displaced towards the east of the continent. This is once more in rough agreement with figure 10.4. However, the anticyclone over North America is found displaced towards the west rather than the east. The reason for this is the effect of the Rocky Mountains which not only alters the relative position of the heat sources and sinks but also compresses the atmospheric column, as discussed in Section 7.4, to produce anticyclonic vorticity over the mountains themselves. Downstream of the Rocky Mountains the stretching of vortex tubes causes the formation of a trough in the upper air flow. Recent computer models have shown that the climatological atmospheric flow is determined by both large mountain barriers and by the distribution of heat sources and sinks. The upper tropospheric flow is most sensitive to the distribution of topography whilst the flow in the lower troposphere is determined by the pattern of continental and oceanic heating. However, the presence of mountain barriers, such as the Himalayas, impede the transport of moisture in the lower troposphere and thus indirectly influence the distribution of heat sources.

In tropical latitudes the differential heating between the Asian land mass and the relatively cool Indian Ocean is responsible for the energy of the south-west monsoon circulation. These monsoon circulations are associated with a large-scale, interhemispheric transfer of latent heat from the southern Indian Ocean and Arabian Sea which, when released over the Indian sub-continent, strengthens the monsoon circulation. The vertical flux of latent and sensible heat from the Arabian Sea during the onset of the south-west monsoon in June has a typical value of $200 \, \text{W m}^{-2}$. This large heat loss produces a rapid cooling of the ocean mixed layer of $2\,^{\circ}\text{C}$ in 1 month and a deepening of the oceanic mixed layer from 40 m in early June to 100 m in early July.

All of the above examples illustrate the complex interconnections which exist between the ocean and atmosphere. They show how the ocean is able to store vast quantities of heat during the inactive season and then release it into the atmosphere, when required, to enhance and to ultimately drive the atmosphere on a variety of scales, from the mesoscale and synoptic-scale circulations to the large-scale monsoon circulations.

10.2. Seasonal anomalies of the ocean–atmosphere system

The ocean plays a major role in the seasonal changes of the atmospheric circulation both in the tropics and at higher latitudes. It is therefore reasonable to assume that anomalies from the seasonal cycle may be linked to anomalous air–sea interactions. Before considering the possible mechanisms of these interactions, an example of anomalous behaviour of the atmosphere and ocean in the North Atlantic Ocean between January and June 1972 will be considered. Figure 10.5 shows both the atmosphere and sea surface temperature anomalies for June 1972. Both patterns are representative of the chosen period between late winter and early summer. The anomaly patterns are obtained from the differences between the observed pattern and the long-term climatic normals for June. The most notable features are:

(i) The negative temperature anomalies, up to $3\,^{\circ}\text{C}$ in the northern North Atlantic, and the positive temperature anomalies, up to $5\,^{\circ}\text{C}$ in the western and central Atlantic at $40\,^{\circ}\text{N}$, as well as smaller negative temperature anomalies at lower latitudes.

(ii) An anomalous cyclonic circulation over the British Isles, with a cold north-westerly flow over the northern North Atlantic, and an anomalous anticyclonic circulation over the central Atlantic Ocean. The cold north-westerly winds mentioned above produced a cold wet spring and early summer in north-western Europe and were associated with the largest volume of ice recorded in the Labrador Current since 1912, when the *Titanic* sank. In June 1972 an extremely active hurricane, called Agnes, followed an unusual northerly path along the east coast of the U.S.A. into the Great Lakes region. It is thought that the anomalous southerly wind off the east coast of the U.S.A. and the high sea surface temperatures prevailing there were responsible for the behaviour of Hurricane Agnes.

From figure 10.5 it can also be seen that the patterns of sea surface temperature anomalies and atmospheric circulation have similar horizontal scales and they are clearly related. The cold northerly atmospheric flow is associated with negative sea surface temperature anomalies whilst the warm southerly flow to the west of the anticyclone occurs over the positive sea surface temperature anomaly. Thus the atmospheric flow is tending to maintain the sea surface temperature anomaly patterns by extracting more heat from the northern North Atlantic Ocean than is normal and by extracting less heat than usual from the western Atlantic. In turn, the large heat flux from the ocean in the northern North Atlantic Ocean would provide an anomalous source of heating which would tend to maintain a low-pressure system downstream of the heat source. Conversely, the reduced heat flux in the western Atlantic would produce an anomalous heat sink which would maintain an anticyclone in the central Atlantic. These arguments are not necessarily conclusive because large-scale interactions with the tropical atmosphere and the Pacific Ocean may also be responsible for the anomaly pattern, but they do suggest that large-scale air—sea interaction can produce a persistent, anomalous pattern which may be reinforced by heat exchange between the ocean and the atmosphere.

In the above example the ocean is behaving as a slave to the atmosphere but, because of its heat storage capacity, it is able to maintain anomalous patterns of heating and cooling. Without the ocean the atmosphere would lose all of its memory of the thermal anomalies within a period of a few weeks. The longevity of the above type of air—sea interaction has aroused much interest in the possibility of using these patterns for long-range seasonal forecasting. However, progress in this area has been limited. Recent studies have shown that the predictability of these air—sea interaction events is limited by the predictability of the atmosphere—which is generally less than 2 weeks. However, once an air—sea interaction event has developed the ocean will tend to maintain the pattern for a period of one to a few months. Thus the atmosphere initiates the anomaly whilst the ocean tends to make it persist. Ultimately the seasonal cycle limits the longevity of these air—sea events by changing the distribution of heat sources and sinks. This eventually results in a breakdown of the anomaly pattern.

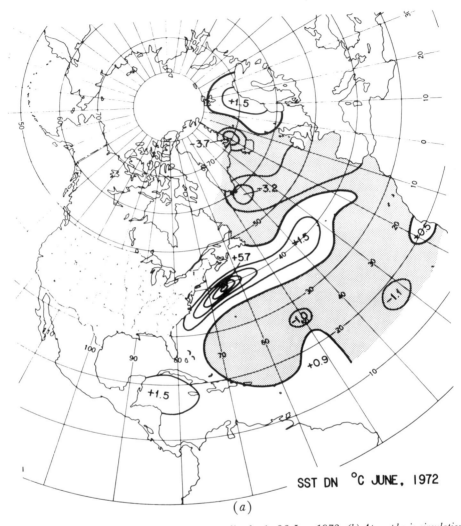

Figure 10.5. (a) Sea surface temperature anomalies for 1–26 June 1972. (b) Atmospheric circulation anomalies at 700 mb height for June 1972. Intervals are in units of 1 standard deviation; maxima and minima are labelled. Zero isopleth is the heavy line. (Reproduced from J. Namias (1973), Q. J. Roy. Met. Soc., **99**(421), 517 and 509, figures 10 and 3, with permission of the Royal Meteorological Society.)

An interesting example of the effect of the seasonal cycle on air–sea interaction events has been described by Namias (1971). In the north-eastern Pacific Ocean it has been found that warm sea surface temperature anomalies, which develop in the summer months, tend to persist through the autumn and into winter. During the autumn, the extra-tropical cyclones intensify in response to the increase in the

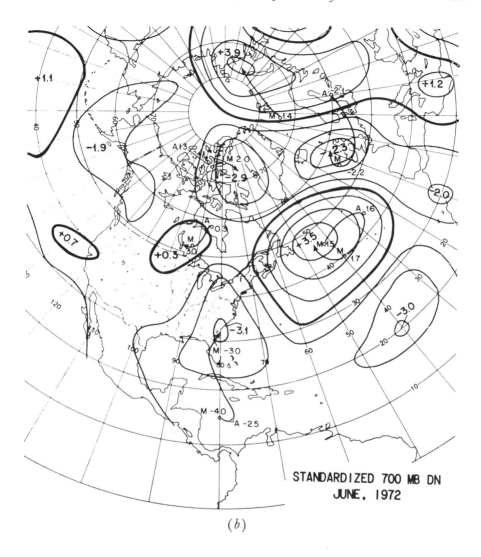

STANDARDIZED 700 MB DN
JUNE, 1972

(*b*)

poleward temperature gradient and the differential heating between ocean and continent. The warm sea surface temperature anomaly increases the heat available, via latent heat and sensible heat fluxes. Thus the energies of the storms are also increased and the cyclones tend to become more intense than is normal during the autumn and winter months. This type of air–sea interaction is, in the end, self-destructive because, once the anomalous supply of heat in the ocean mixed layer has been exhausted, the anomalous atmosphere circulation can no longer be maintained. However, it does illustrate the point that the amount of heat stored in the ocean mixed layer during the summer heating season is an important factor for the prediction of air–sea interaction events in the subsequent autumn and winter seasons. A summertime sea surface temperature anomaly of 1°C, extending to a depth of 50 m, contains approximately 2×10^8 J m^{-2} of excess energy. If this energy is released over the subsequent 3 months, it will provide an anomalous heat source of about 25 W m^{-2}, compared with the normal autumnal heat loss of 80 W m^{-2}, in the north-eastern Pacific.

There is also evidence that positive sea surface temperature anomalies in the Arabian Sea, which develop in the heating season prior to the onset of the summer south-west monsoon, may provide an additional source of latent heat to enhance the monsoonal circulation. In the tropical ocean relatively small sea surface temperature anomalies of approximately 1°C, can produce large anomalies in evaporation because of the non-linearity of the water-vapour pressure with temperature.

In all of the above examples, the interaction of the atmosphere with the ocean has been confined to the mixed layer which occupies the upper 50–200 m of the ocean. As seen in Section 5.2, most of the seasonal storage of heat is confined to this layer and therefore many short-term or seasonal anomalies can be explained by variations in the depth and temperature of the mixed layer. However, other interactions may also be significant. The ocean transport of heat shows a seasonal variation both in tropical latitudes and middle latitudes, as illustrated by figure 5.3. In middle latitudes the seasonal transport variation is small in comparison with the storage term but in the tropics it is larger than the storage term. These seasonal changes in the ocean heat transport, at low latitudes, are related to the seasonal variations in the trade winds, with strong trade winds enhancing the poleward transport of heat. Model studies have shown that this seasonal response of the ocean is confined to the direct stress-driven flow in the mixed layer, via the Ekman balance. Hence strong westward trade winds will produce a poleward transport of warm surface water and colder sub-surface water returns towards the equator for mass balance. In middle latitudes the seasonal variations in heat balance are not well understood and will not be considered in detail here.

10.3. *Interannual fluctuations in the ocean–atmosphere system*

The air–sea interaction events discussed in the previous section occur on monthly and seasonal time scales. They are generally connected with the short-term variability of the atmosphere or the seasonal forcing of the atmosphere–ocean system. On the interannual time scale there is no large external forcing and therefore variations arise from internal interactions within the Earth's ocean–atmosphere system. The most significant of these internal interactions is that of the El Niño–Southern Oscillation (ENSO) phenomenon. Two or three times each decade anomalously warm water, perhaps $2-4°C$ above normal, appears off the coast of Peru and Ecuador and persists for a number of seasons. The event usually occurs in December or January, hence the name El Niño, or Christchild. Normally the Peruvian coast is a region of strong coastal upwelling. The upwelling brings large quantities of nutrients to the sea surface and this maintains a very productive anchovy fishery. However, during El Niño warm equatorial waters, which are low in nutrients, move southwards along the Peruvian coast and replace the cold, nutrient-enriched waters. The anchovy industry is drastically affected when this occurs. After the 1972 El Niño the fishery industry was virtually eliminated and, to this day, stocks have not recovered to their pre-1972 values.

Though one cannot overestimate the importance of El Niño to the fishing economies of Peru and Ecuador, it has only recently been realized that El Niño is just one symptom of a world-wide ocean–atmospheric event. The oceanic component of the event is located in the eastern, tropical Pacific Ocean. Figure 10.6 shows the sea surface temperature anomaly in December 1982. This was associated with one of the largest El Niño events in the last 100 years. It can be seen that the anomaly extends from the South American coast westwards to beyond the International Date Line and from $10°N$ to about $20°S$. This is an area equivalent to about 10% of the global ocean surface. If the average anomaly temperature is taken to be $2°C$, which is a conservative estimate, and if it is assumed that the depth of the anomaly is confined to the upper 100 m of the ocean, then the excess heat stored is 3×10^{22} J. This is equivalent to about eight times the total available potential and kinetic energy of the atmosphere. Furthermore, if all this anomalous heat was converted into available potential energy at the observed rate of $2 \cdot 5 \ \mathrm{W\,m^{-2}}$, then it would be sufficient to maintain the general circulation of the atmosphere for about 9 months. This is an overestimate, of course, because an energy conversion efficiency of 100% has been assumed whilst in reality a proportion of the heat will be lost by long-wave emission to space. It does, however, illustrate the vast quantity of anomalous energy associated with the ENSO phenomena.

Before discussing the changes in atmospheric circulation associated with El Niño, it is useful to describe the salient features of the normal circulation over the tropical Pacific Ocean during the Northern Hemisphere winter. This circulation is shown in figure 6.8. The surface circulation is controlled by the position of the North Pacific and South Pacific sub-tropical highs and the low pressure over the maritime continent of Indonesia. The trade winds flow out of the high-pressure regions and converge in two main zones:

(i) The ITCZ which is located about $5-10°N$ of the equator and extends from Central America to Indonesia.

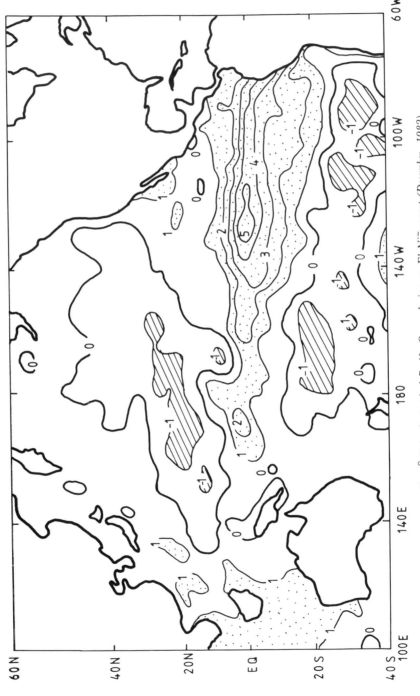

Figure 10.6. Sea surface temperature anomalies (°C) in the tropical Pacific Ocean during an El Niño event (December, 1982).

(ii) The South Pacific convergence zone (SPCZ) shown in figure 6.8, which extends from the Indonesian region south-eastwards into the central South Pacific.

To conserve mass in the convergence zones it is necessary to have an outflow of air in the upper troposphere directed towards the sub-tropical highs. The north–south component of this circulation is the Hadley cell whilst the east–west component is known as the Walker circulation. The convergence zones are associated with heavy precipitation and large latent heat release. They therefore provide the kinetic energy to drive the Hadley and Walker circulations. The region between the two convergence zones, in the tropical eastern Pacific Ocean, is an area of low precipitation, hence its label of 'dry zone', and low sea surface temperature. The cold sea surface temperatures are maintained by the strong south-east trade winds which cause pronounced upwelling along the equator and along the South American coast. This area of relatively cold water, having sea surface temperatures of $18–24°C$, is known as the cold tongue.

During an El Niño event the cold-tongue zone is replaced by warmer water from the central part of the Pacific Ocean. In an extreme El Niño event, such as occurred in 1982–1983, the warming is sufficient to completely remove the east–west sea surface temperature gradient along the equator. The normal atmospheric circulation pattern changes in a number of distinct ways during such an event. First, the South Pacific High is generally much weaker than normal and therefore the Southern Hemisphere Hadley–Walker circulation is also weaker. In the western and central Pacific the normal trade winds are replaced by westerly winds. The surface pressure over Indonesia and the Indian Ocean also becomes anomalously high. This anomaly in surface pressure between the south-eastern Pacific and the Indian Ocean is known as the Southern Oscillation and it was first discussed by the British meteorologist Sir Gilbert Walker early this century. The second major change in the atmospheric circulation is that the main convergence zones are displaced. The ITCZ moves southwards towards the equator by between 5 and $10°$ of latitude whilst the SPCZ moves eastwards. The region of heaviest precipitation is displaced from the Indonesian and western Pacific sector to the central Pacific and islands which are normally in the dry zone may experience very heavy rainfall. For example, the rainfall at Christmas Island $(0°, 160°W)$ between August 1982 and January 1983 was 228 cm, which may be compared with the normal annual rainfall of 20 cm. At the same time the Indonesian sector, including the Australian continent, experienced long periods of drought due to the anomalous high pressure. In extreme events these droughts may last for up to 2 years.

The effect of the ENSO interaction is not restricted to the tropical

Pacific zones. The changes in the tropical pressure distribution, associated with the Southern Oscillation, has a global impact on the general atmospheric circulation and affects both monsoonal circulations over the Asian and African continents as well as the extra-tropical zones. Figure 10.7 shows the anomalous influence of a typical El Niño event on the middle-latitude circulation over the North Pacific Ocean and over the North American continent. The tropical heat source produces a train of Rossby waves which disrupt the normal circulation patterns. It also produces an anomalous low-pressure area over the Aleutian Islands and an anomalous high-pressure area over western North America. It will be noted that the influence on higher latitudes is much stronger in the winter hemisphere than in the summer hemisphere.

The causes and maintenance of the ENSO phenomenon by air–sea interaction processes are not well understood. Although some scientists have suggested that the phenomenon may be predicted up to 1 year in advance, there is no evidence to date that these predictions will be

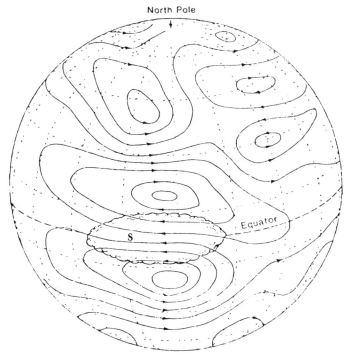

Figure 10.7. Atmospheric circulation anomalies over the Pacific Ocean for December 1982 to February 1983. **S** *denotes the equatorial heat source caused by the SST anomaly, it is associated with anomalous convective cloud and heavy rainfall. (Reproduced from M. A. Cane (1983),* Science, **222**, *1201, figure 6, with permission of the American Association for the Advancement of Science.)*

sufficiently reliable to be of practical value. However, during the 1982–1983 El Niño the progression of events was well observed both in the atmosphere and the ocean. The main features of the development of the El Niño were:

(i) A decrease in the Southern Oscillation index, associated with the collapse of the South Pacific High, which preceded all other developments.

(ii) An eastward progression of the sea surface temperature anomaly from the central to the eastern Pacific where it reached its maximum amplitude.

(iii) An eastward progression of the region of heaviest precipitation from the western to the central Pacific Ocean.

(iv) An increase in the depth of the equatorial thermocline in the eastern Pacific from 50 to over 200 m.

(v) The growing predominance of westerly winds extending from the western to the central Pacific.

At the maximum amplitude of the El Niño there was little or no tilt of the thermocline and the equatorial undercurrent disappeared for 2–3 months. From all of the above observations some general conclusions can be drawn. First, the collapse of the South Pacific High, which precedes all of the other events, appears to be initiated by the atmosphere like other, shorter time scale, air–sea interactions. However, over the equatorial Pacific the ocean and atmosphere tend to amplify each other's responses. The weakening of the south-east trade winds on the equator will reduce the westward tilt of the equatorial thermocline which, in turn, will move the sub-surface cold water in the eastern Pacific further away from the surface. The large net heat flux, between 50 and 120 W m^{-2}, into the eastern Pacific Ocean implies that the mixed layer will, in the absence of cold upwelling, warm steadily by about 2 °C in 3 months over a depth of 100 m. Some additional warming will occur by advection of warmer water from the central Pacific to the eastern Pacific when the thermocline relaxes. The increase in sea surface temperature will, in turn, move the region of highest sea surface temperature from the western to the central Pacific. The convergence zones tend to follow the region of highest sea surface temperature and so they will be displaced eastwards into the central Pacific Ocean. The surface wind circulation over the equatorial zone is therefore crucial in the air–sea interaction loop. It changes the slope of the thermocline and alters the upwelling. The subsequent changes in sea surface temperatures affect the wind circulation and the interaction cycle begins again. The whole progression of events takes between 1 and 2 years, on average, from the reduction in the South Pacific High to the final decay of the sea surface temperature anomalies.

10.4. Long-term variations in the ocean–atmosphere system

In order to understand the present climate and possible future changes in the climate, it is useful to consider the record of past climate. Figure 10.8 shows a reconstruction of the Earth's surface temperature during the past 1 million years. It is necessary to remind the reader that the basis of this reconstruction is not as rigorous as present-day measurements of climate and, of course, it can never be so. Only the last 200 years of the climate record is based on direct measurements whilst the remainder has been deduced from proxy data. This type of data can assume many forms. For example, the width of tree rings are known to be correlated with droughts and abnormally cold years. Tree ring widths can therefore be used to deduce climatic variations over the last 1000 years. Historical records of variations in alpine glaciers and in continental ice sheets have also been used to deduce relatively recent variations in climate. From studies of pollen in lake sediments and peat bogs paleobotanists are able to ascertain past changes in surface vegetation which can, in turn, be related to major shifts in climatic zones. However, the analysis of deep-sea ocean cores obtained in the last two decades from the deep-sea drilling programme (DSDP) have provided the scientific community with the most comprehensive record of long-term climate variations. The ratio of the heavy isotope of oxygen, ^{18}O, to ordinary oxygen, ^{16}O, found in the calcite shells of marine animals, is dependent on the prevailing temperature during the creature's lifetime. By selecting a species which is known to exist either in the surface layers of the ocean, or in the benthic layers, it is possible to deduce the temperature of the ocean when the organism was laid down in the sediment. By this technique an ocean sediment core can give a record of both surface and bottom temperature from the last Ice Age to about 50 million years ago. The oxygen ratio in benthic foraminfera can also be used to deduce the total volume of ice locked up in the continental ice caps, as shown in figure 10.8. Cores from many different locations have shown a similar oxygen isotope ratio variation with time. As it is unlikely that, even in the Ice Ages, the bottom temperatures were greatly different from their present values of $2–3°C$, the isotopic variations are most likely to be associated with changes in the isotope ratio. As large amounts of water are removed from the ocean during glaciations, the water column becomes enriched with the heavier ^{18}O isotope. Thus the changes in the isotope ratio obtained from core analysis can be used to obtain a direct estimate of the amount of water lost from the ocean and deposited in ice sheets. More recent climatic variations have been deduced from the analysis of ice cores from the Greenland and Antarctic ice sheets which reveal distinct layering as the result of annual variations in snowfall. The $^{18}O{:}^{16}O$ ratio in these layers is dependent on the temperature at the time of the formation of the snow

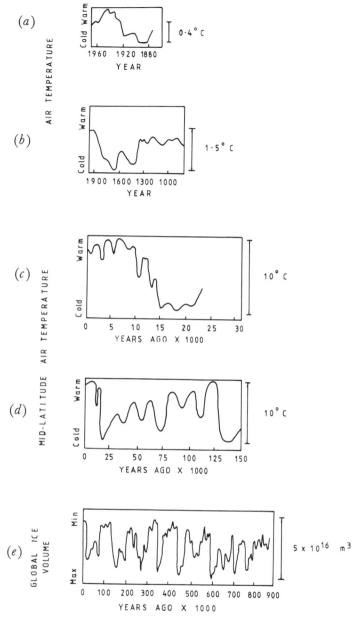

Figure 10.8. General trends in global climate; the past million years. (a) Changes in the 5 year average surface temperatures over the region 0–80°N. (b) Winter severity index for eastern Europe. (c) Generalized Northern Hemisphere air temperature trends, based on fluctuations in alpine glaciers, changes in tree lines, marginal fluctuations in continental glaciers and shifts in vegetation patterns record-ed in pollen spectra. (d) Generalized Northern Hemisphere air temperature trends based on mid-latitude sea surface temperature, pollen records and world-wide sea level records. (e) Fluctuations in global ice volume recorded as changes in isotopic composition of fossil plankton in a deep-sea core. (Reproduced from The Physical Basis of Climate and Climate Modelling, *WMO Garp Series, No. 16, p. 8, figure 2.5, with permission of WMO.)*

and it can therefore be used to obtain an annual temperature recording extending over the past 10,000 years. Furthermore, air bubbles trapped in the ice core can give useful information on the past variations of atmospheric composition, especially those of carbon dioxide. Ash deposits from volcanic eruptions are also faithfully recorded in the ice.

The record of temperature shown in figure 10.8 shows that the variations in the recent past are about one order of magnitude less than those associated with the glacial and interglacial cycles. In the last century the most notable feature is the warming of $0 \cdot 6°C$ which took place between 1880 and 1940. This change, although small, was associated with marked changes in climate. The circumpolar vortex contracted, the planetary Rossby waves decreased in amplitude and there was a world-wide retreat of mountain glaciers. The amplitude of the warming was higher in the Atlantic sector of the Arctic and this caused a poleward contraction of the sea ice. Since 1940 the global trend has been towards a steady cooling which is most pronounced in the Arctic. There is evidence for an increased amplitude of the Rossby waves and this is associated with more extreme weather than that observed in the warm period earlier this century. However, it is necessary to point out that variations in regional climate may not necessarily follow the global trend in temperature. During the period of global warming, the middle latitude zone of Eurasia showed a cooling of $0 \cdot 5°C$.

During the last 1000 years a period of generally cooler climate, known as the Little Ice Age, prevailed between 1450 and 1850. The climate was at its most extreme in the late seventeenth century and a major expansion of mountain glaciers occurred, especially in the continental areas adjacent to the North Atlantic Ocean. Figure 10.9 shows the ocean-temperature anomaly pattern between 1780 and 1820, measured relative to the temperatures in the early twentieth century. Generally higher temperatures prevailed in the sub-tropical Atlantic Ocean whilst there were lower temperatures in the area of Iceland. The winters of this period were characterized by a weaker Icelandic low-pressure system, a reduction in the westerly winds over Europe and by more frequent 'blocking' of the westerly flow. There is evidence for an equatorward displacement of the climatic zones by between 2 and 4° of latitude and for an increase in the pack ice in the Arctic. The equatorial displacement of the depression tracks resulted in the more frequent occurrence of winter cyclonic storms in the Mediterranean region. Prior to the Little Ice Age, between 1000 and 1300, the climate was in a more stable state with temperatures similar to those of the present day.

The record from 1000 to 20,000 years ago shows the approach of the last Ice Age. Early in the period, about 5000–8000 years ago, the climate reached an optimum and global temperatures were $1–2°C$ higher than at present. Geological evidence has shown that during this period the mean

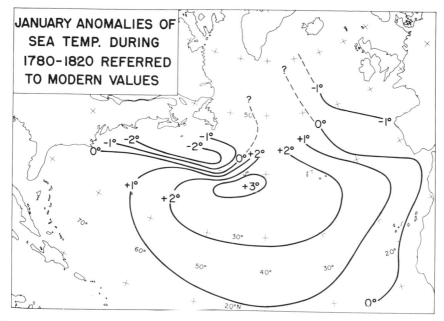

Figure 10.9. January 1780–1820 anomalies of sea temperatures referred to modern normals. (Reproduced from H. H. Lamb and A. I. Johnson (1969), Geografiska Annaler, **94**, *123, figure 13a, with permission of Almqvist and Wiksell.)*

sea level was generally higher, and therefore ice caps were rather smaller, than at present. The climate was wetter than now, especially over the sub-tropical Saharan and north-western Indian regions, and it is probable that the monsoons were more intense. It is noteworthy that the initial development of civilization took place in this favourable climatic period.

On the longer time scale the most notable features are the large oscillations in global temperature which are associated with the glacial cycles. The periods of these cycles vary from 120,000 to 40,000 years. The warmest periods of these cycles, comparable with the present-day climate, lasted about 10,000 years. Statistically, our present interglacial should end in the next 1000 or 2000 years. During the interglacial periods the ice has been confined to the Antarctic and Greenland ice caps whilst in the glacial periods continental ice sheets covered large areas of the northern continental land masses. At the glacial maximum about 9% of the Earth's surface was covered with ice sheets. This may be compared with about 3% ice cover during interglacial periods. The large volume of water, typically 75×10^6 km^3, bound up in ice sheets during a glacial period implies a drop of 100 m in the mean sea level. This would expose large areas of the continental shelf. Figure 10.10 shows the surface conditions occurring during the last Ice Age, 18,000 years ago. It can be seen

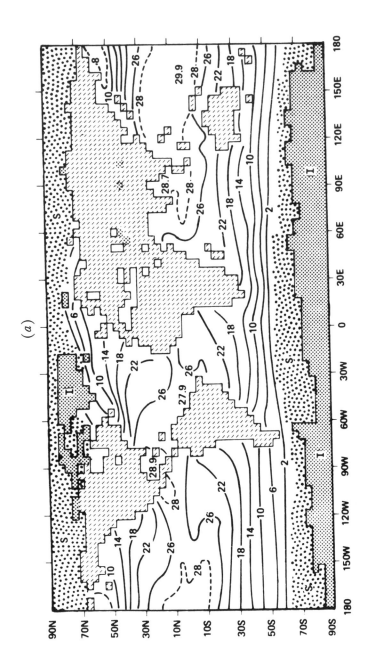

(a)

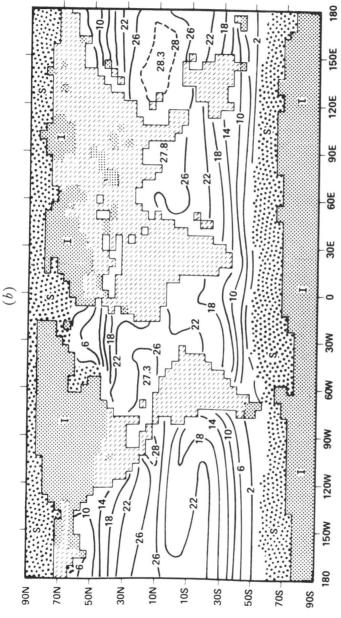

Figure 10.10. (a) Present annual mean sea surface temperature. (b) Estimated sea surface temperature 18,000 years ago at height of glaciation as determined by CLIMAP studies of fossil assemblages in deep-sea cores. (Reproduced from W. L. Gates (1976), Modelling the Ice Age climate, Science, 191, 1139, with permission of the American Association for Advancement of Science.)

that although there are large reductions in ocean surface temperature in polar latitudes, in the sub-tropical and equatorial zones the sea surface temperatures are only $1-2°C$ lower than at present. The increased surface temperature gradient between the tropics and polar latitudes would have increased the zonal atmospheric circulation both in middle and lower latitudes. This may, in turn, have increased the circulation of the large sub-tropical ocean gyres. However, the sea surface temperature pattern in the North Atlantic Ocean shows evidence that most of the Gulf Stream water was directed eastwards towards southern Europe and that the north-eastern Atlantic drift current was weaker than it is today. It has also been suggested, though not proven, that the production of bottom water in the northern North Atlantic Ocean was less during the Ice Age because of the presence of a halocline in the sub-polar gyre. This reduction in bottom-water production would have weakened the meridional thermohaline circulation and reduced the transport of heat into the sub-polar gyre.

In the eastern equatorial Pacific Ocean the cooling was much larger than in other regions of the tropical ocean. This is indicative of enhanced upwelling and the strengthening of the south-east trade winds during the glacial period. In tropical latitudes the monsoons were weaker than at present and generally drier conditions prevailed over the land masses in this region.

It was mentioned earlier that the temperature of the ocean bottom water inferred from benthic foraminfera appears to have changed little during the glacial cycles. The present-day deep ocean is therefore a glacial ocean and is a symptom of a much longer Ice Age which began over 2 million years ago. Recent evidence from deep-sea cores has shown that the bottom water may have been about $12°C$ warmer 45 million years ago than it is today. The bottom waters then cooled progressively reaching a temperature of $2°C$ about 15 million years ago. It is thought that this cooling may be related to changes in the distribution of oceans and continents and especially to the displacement of Antarctica to its present polar position and to the confinement of the Arctic Basin.

10.5. Mechanisms of climate change

The large variation in amplitude and time scales of the climatic record, depicted in figure 10.8, suggests that it is highly unlikely that one mechanism is responsible for all climate fluctuations. An idea of the variety of potential processes which may contribute to these fluctuations can be obtained from figure 10.11. These potential processes can be broadly grouped into external and internal mechanisms. The main external mechanisms include changes in energy output from the sun and

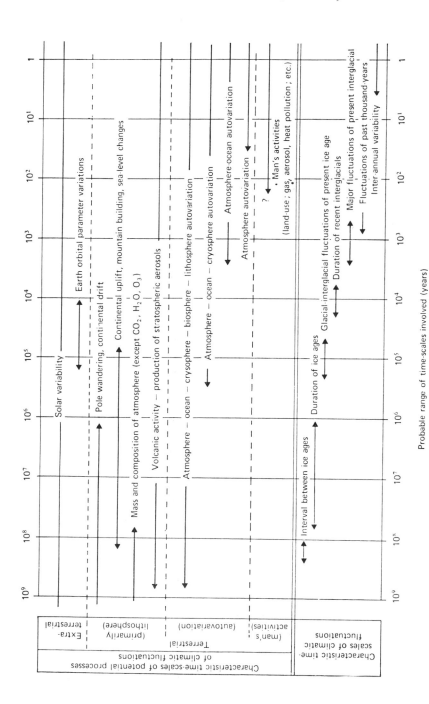

Probable range of time-scales involved (years)

Figure 10.11. Examples of potential processes involved in climatic fluctuations (top) and characteristic time scales of observed climatic fluctuations (bottom) (Kutzbach, 1974). (Reproduced from The Physical Basis of Climate and Climate Modelling, WMO Garp Series, No. 16, p. 9, figure 2.6, with permission of WMO.)

variations in the Earth's oribtal parameters. Changes in the solar output would have a direct effect on the Earth's radiative temperature, as shown by equation 1.4. For example, a postulated 1% change in the solar constant would change the Earth's radiative temperature by $0 \cdot 6°C$. Long-term variations in the Earth's orbital parameters would change the seasonal distribution of radiation but would not alter the total radiation incident upon the Earth.

The internal mechanisms can be sub-divided into three groups. First, there are the processes which involve the Earth's lithosphere such as continental drift, mountain building and volcanic activity. Second, there are the internal oscillations between the atmosphere, ocean, cryosphere, biomass and land surface. For example, the ENSO phenomenon has already been shown to involve an internal interaction between the ocean surface and the atmosphere. Thirdly, there are the effects of man's activities which may have begun to affect the climate system by the addition of carbon dioxide to the atmosphere through the burning of fossil fuels, by adding aerosols to the atmosphere and by changes in land use, particularly deforestation.

In view of the large number of potential mechanisms it is perhaps not so surprising that so many misconceptions concerning climate change have arisen in the popular literature. For example, from figure 10.11, six different processes may be identified as possible causes for the glacial and interglacial cycles. It is possible that all of these six processes contribute to the cycle. The result of all of the interactions, each with different time scales, may produce such a complex response that it is not possible to distinguish any individual process or to separate cause from effect. The complexity of the climate system has led to the development of a number of different approaches by climatologists in their efforts to achieve a greater understanding. Two methods will be considered here. Both have been applied to the glacial–interglacial cycles. A popular theory for the explanation of the glacial cycles is the astronomical theory of orbital changes which was developed by Milankovitch (1938). Milankovitch proposed that Ice Ages were initiated in the high latitudes of the Northern Hemisphere by a sequence of cold summers which would allow an accumulation of ice to form from one winter to another. The increase in albedo of high latitudes would, in turn, cool the surface and produce a positive feedback mechanism. He further reasoned that variations in the obliquity, or tilt, of the Earth, the eccentricity of the Earth's orbit and the precession of the equinoxes may occasionally result in a significant depletion of solar radiation during the high-latitude summer. By calcula-tion of the orbits of the Earth, he showed that the obliquity had a period of 20,000 years, the precession of the equinoxes had a period of 40,000 years and that the eccentricity had a period of 100,000 years. The sum of all these processes was shown to cause variations of 12% in the monthly solar radiation. Spectral analysis of the oxygen isotope record from deep-sea cores has shown that similar periods to those predicted by the astronomical theory have occurred in the past. It is also possible to use the astronomical theory to calculate variations in the incident solar radiation from the present day to several million years ago and to make direct comparisons with the isotope record. These calculations have shown that good statistical fits of the predictions to the experimental data can be obtained (figure 10.12).

A second approach to the problem is to develop a model of the climate system based on the essential physics thought necessary to explain the glacial–interglacial phenomenon. A basic model for glacial cycles would have to include an atmosphere, an ocean, an ice sheet and isostatic adjustment of the lithosphere to the ice sheet. A set of differential equations for each of the components of the climate system would then be solved to deduce either the natural oscillation of the system or its response to

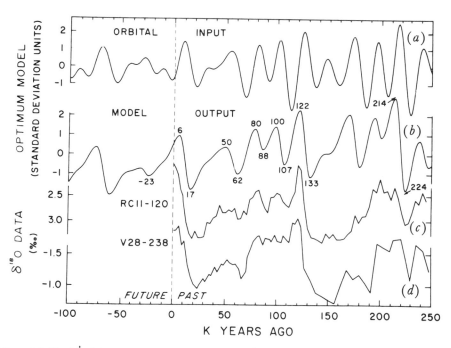

Figure 10.12. (a) Solar radiation for summer at 65°N due to long-term variations in orbital parameters. (b) Output of a climate model, with a time constant of 17,000 years, and allowance for glacial growth and decay. According to this model, the influence of orbital variations over the next 23,000 years will be to enlarge continental ice sheets. (c) Oxygen isotope curve for deep-sea core RC11-120 from the southern Indian Ocean. (d) Oxygen isotope curve for deep-sea core V28-238 from the Pacific Ocean. (Reproduced from J. Imbrie, p. 535, figure 3, in Climatic Variations and Variability: Facts and Theories, *edited by A. Berger, with permission of Reidel.)*

forcing, for instance by Milankovitch cycles of solar radiation. Figure 10.13 shows the natural oscillation of a relatively simple Ice Age model. The periods of oscillation are controlled by the time scale of the build-up, or decay, of the ice sheets, which is 5000 years, and by the overturning time scale of the ocean which is 1000 years. Auto-oscillations are possible because the time scales of the two systems are similar. The interesting result from this model is the phase relationship between the ocean surface temperature, T, and the area of the ice cap, l. The maximum ocean temperature precedes the maximum extent of the ice cap by approximately 2500 years. The high ocean temperature is necessary to provide enough precipitation to build up the ice cap. The subsequent increase in albedo of the ice cap causes a reduction in the ocean temperature. This drop in ocean temperature eventually causes a reduction in the rate of precipitation which then results in the ablation of the ice cap and a decrease in the albedo. The interaction of these two processes can produce glaciation cycles with periods of 5000 years and temperature amplitudes of $10°C$ without invoking external influences. However, the natural period is too short to explain many of the long-term glaciation periodicities. These types of models have also been used to estimate the response to the Milankovitch cycles of solar radiation. Some model studies have found that the magnitude of the solar radiation anomalies is insufficient to cause the large amplitude temperature variations of the

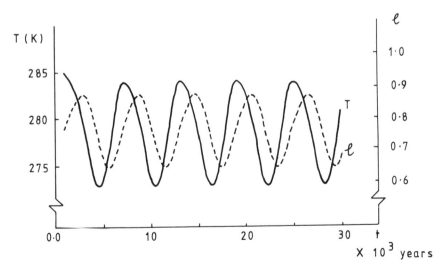

Figure 10.13. An example of self-sustained, periodic oscillations of the climatic system (after Kallen et al., 1979). The ocean surface temperature, T, is shown by a solid curve and the ice sheet extent, l, by a dashed curve. (Reproduced from E. Kallen et al., 1979, J.A.S., 36, 2299, figure 6b, with permission of the American Meteorological Society.)

glacial cycles and that random perturbations in the internal system may be just as effective in initiating glacial cycles.

Climate models have also been used to investigate more recent fluctuations in climate such as the warming trend between 1880 and 1940. Three mechanisms which have been investigated are:

(i) Volcanic eruptions, which change the stratospheric aerosol.
(ii) Variations in solar radiation associated with sunspot cycles.
(iii) The increase in anthropogenic carbon dioxide.

The general conclusions from these experiments are that solar radiation variability does not account for the warming trend but that both volcanic aerosol and carbon dioxide can produce changes in global surface temperature similar to those observed. Figure 10.14 shows the variation in volcanic aerosol during the last 100 years and its relation to the surface temperature. It can be seen that the absence of major volcanic eruptions earlier this century should have increased the net solar radiation reaching the ground assuming that other factors, such as cloud cover of the globe, did not change. More detailed radiation models have shown that a volcanic dust cloud, typical of a major eruption, in the stratosphere will produce a net cooling at the Earth's surface of a few tenths of a degree. Furthermore, because of the thermal inertia of the ocean, this cooling effect may persist for 1 or 2 years after the disappearance of the volcanic dust cloud.

The effect of an increase in atmospheric carbon dioxide is a current research problem, concerning many environmental scientists. The change in temperature from 1880 to the present day is estimated from computer models to be approximately $0 \cdot 3°C$. Figure 10.15 shows the projected trend in carbon dioxide emissions and the consequent calculated impact on surface temperatures for the next 50 years. These analyses are based on assumptions about the rate of growth of the world economy. For a high rate of growth the surface temperature is predicted to rise by $1 \cdot 4$ K by 2030 whilst for a low rate of growth the rise is closer to 1 K. However, it is clear that both of these temperature increases are larger than natural variations in mean temperature occurring over the last 500–1000 years and, if correct, they suggest that significant movement of climate zones may take place in the next 50 years. The imperfections in our understanding of the climate system imply that a considerable research effort is required to verify or disprove these predictions.

These examples have shown that by the careful use of statistical techniques and of physical models of the climate system, it is possible to investigate some of the causes of climatic change. As was shown with the astronomical theory of climate, the results are not necessarily definitive

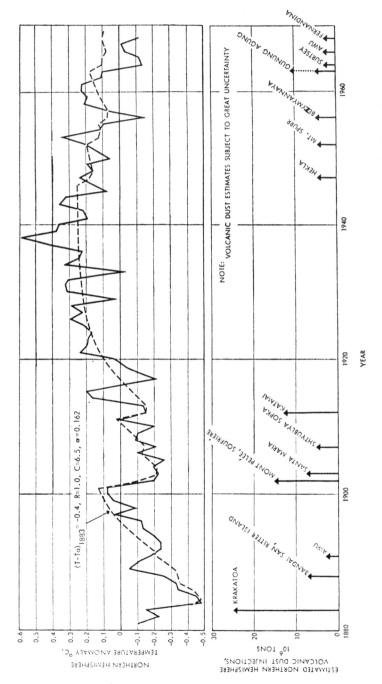

Figure 10.14. Variation of mean surface temperature of the Northern Hemisphere since 1881 (solid curve), and estimate of temperature changes resulting from known volcanic eruptions (dashed curve). Lower curve shows estimated volcanic dust injections in the Northern Hemisphere from known eruptions. (Reproduced from J. M. Mitchell, p. 130, figure 2, in Physical Basis of Climate, WMO Garp Series, No. 16, with permission of WMO.)

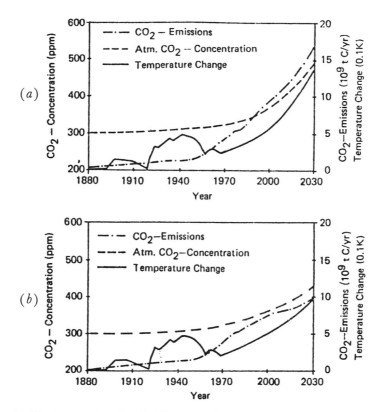

Figure 10.15. Atmospheric carbon dioxide concentration and global temperature change for (a) high and (b) low projected growth in consumption of fossil fuels. (Reproduced from F. Niehaus, p. 658, figure 12, in Climatic Variations and Variability: Facts and Theories, *edited by A. Berger, with permission of Reidel.)*

but careful interpretation of climate data and model results can shed considerable light on this very complex climate system.

10.6. Perspectives in ocean and atmospheric science

Both meteorology and oceanography are observational sciences and therefore the growth in understanding of how these complex systems work has been largely dependent on improvements in the methods of measurement. For example, the establishment of the synoptic network of upper air stations using the radiosonde in the 1940s and 1950s provided, for the first time, daily information on the pressure and wind fields in the troposphere and lower stratosphere. These observations led to an improved understanding of the development of weather systems which, in turn, laid the foundations for numerical weather predictions. Similarly in oceanography it has been seen that the neutrally buoyant float provided the first important evidence for mesoscale eddies in the deep ocean as well as some of the first observations of the abyssal western boundary currents. Both of these phenomena had not been previously observed because of the coarse horizontal resolution of hydrographic data.

Despite the considerable improvements in instrumentation, both meteorologists and oceanographers today are confronted with questions which, if they are to be answered satisfactorily, require additional methods of measurement to those discussed previously. One important question involves the role of the ocean in seasonal and interannual variations in climate. In the previous section it was seen that the 1982–1983 ENSO event was associated with global weather variations. It will be necessary to measure both the ocean and the atmosphere, on an ocean basin and global scale, respectively, in order to acquire a fuller understanding of such events. The Tropical Ocean–Global Atmosphere (TOGA) experiment aims to study these phenomena over the next 10–15 years. In order to detect the predicted warming trend due to the effect of anthropogenic carbon dioxide in the atmosphere, it will be necessary to measure relatively small global and regional changes in temperature over decades. Unfortunately, the conventional atmospheric and oceanographic data are not collected in a uniform manner over the globe. The network of radiosonde stations is strongly biased towards the Northern Hemisphere land masses and, apart from the occasional weather ship and island station, the upper air data over the oceans are sparse. Surface marine meteorological data, which form the basis for estimates of surface heat exchange and for evaporation, are confined to the major trade routes. Thus large areas of the ocean, especially the South Pacific and southern Indian Oceans, are devoid of data. Furthermore, although research ships do provide high-quality information, they are insufficient in numbers to provide the horizontal resolution required. Rising fuel costs indicate that the number of merchant ships will not increase and therefore the number of surface observations is unlikely to increase in the near future.

The only forseeable possibility for obtaining global- and basin-scale observations at a reasonable cost is the satellite combined with automatic ocean buoys. A polar-orbiting satellite is able to cover most points on the globe at least once a day, whilst a geostationary satellite, positioned above the equator, is, in principle, able to give continuous monitoring of one section of the Earth. Five geostationary satellites can give complete coverage of the globe. However, because of the curvature of the Earth, the resolution polewards of $50°$ latitude is inferior to that obtained with the polar-orbiting satellites. Everyone is aware of the use of these types of satellites to provide up-to-date pictures of the weather systems in their area of the globe but fewer people may be aware of the equally exciting potential of the satellite for quantitative measurements of the atmosphere and ocean. Satellite systems can be distinguished as passive or active. The former designation refers to a system which measures electromagnetic radiation, either emitted or reflected from the ocean or atmosphere, whilst the latter refers to the use of a controlled source of radiation such as radar or lidar.

Until recently most meteorological satellite systems were passive. They measured the radiation emitted in the infra-red and microwave regions of the spectrum and they also monitored reflected solar radiation, as shown in table 10.1. Passive sensors in the visible are able to delineate regions of clouds, ice boundaries and the colour of the land and sea surfaces during daylight hours. However, thermal infra-red sensors, which measure the radiation temperature of the Earth's surface or cloud tops, are able to delineate variations both during the day and the night. The thermal emission from the sea surface can be used to obtain the sea surface temperature using known values of the emissivity. This sea surface temperature is representative of the upper 100 μm and it is therefore different from the bulk temperature typical of the upper 1–3 m measured by ships. However, except under low wind conditions, the differences are usually less than 1°C. Unfortunately for sea surface temperature measurements, the atmosphere also emits radiation at infra-red wavelengths, due to the presence of carbon dioxide, ozone and water vapour, and therefore the signals have to be corrected for atmospheric effects. Furthermore, clouds will absorb all thermal radiation emitted from the sea surface and therefore the technique is limited to cloud-free regions. Microwave energy is also emitted from the sea surface and, though much less than the emitted infra-red energy, at certain wavelengths, it can be transmitted through non-precipitating clouds and used to obtain sea surface temperature measurements. The thermal emission of radiation by the atmosphere has been used to obtain the vertical distribution of temperature in the atmosphere. The principle of the instrument used is that different wavelengths of infra-red and microwave radiation originate at different levels in the atmosphere. The radiance at each wavelength can therefore be related to the temperature at the appropriate level. The vertical resolution is not as good as that obtained from a radiosonde, being about 5 km, but the horizontal resolution is much better, being typically 100 km for infra-red observation and 250 km for passive microwave measurements.

Passive microwave sensors have been used to determine the total water vapour and the total liquid water content of the atmospheric column from simultaneous measurements of the water vapour absorption line at 1·36 cm wavelength and the liquid water absorption band at 0·97 cm wavelength. From such measurements global maps of water

Table 10.1. Selection of atmospheric and oceanic parameters measured by satellite.

Environmental parameter	Satellite sensor	Accuracy	Method
Tropospheric wind	Visible/infra-red	∓ 4 m s^{-1}; $\mp 20°$	Cloud displacement from geostationary satellite
Surface winds	Scatterometer (radar)	∓ 2 m s^{-1}; $\mp 15°$	Scattering of radar beam by sea surface
Waves	Altimeter (radar)	$\mp 0{\cdot}3$ m	Shape of reflected pulse related to significant wave height
Sea surface elevation	Altimeter (radar)	$\mp 0{\cdot}1$ m	Time delay of radar pulse
Vertical atmospheric temperature profile	Infra-red/passive microwave	∓ 2 K	Multichannel radiometer samples emitted thermal radiation from different levels of atmosphere
Sea surface temperature	Infra-red/passive microwave	$\mp 0{\cdot}5$–1 K	Thermal radiation from sea surface Corrections for atmospheric contamination Infra-red limited to cloud-free areas
Solar radiation at sea surface	Visible/near infra-red	∓ 17 W m^{-2}	Measures cloud-top temperature
Precipitation	Visible/infra-red	Qualitative	Empirically related to precipitation
Precipitation	Passive microwave	100%	Measures liquid water content of atmospheric column
Ice cover	Visible/infra-red Passive microwave	—	High resolution in visible and infra-red Low resolution, but all weather, in microwave

vapour and liquid water content have been produced. These maps would have been unobtainable with the same resolution from conventional observations. Absorption of microwave radiation at a wavelength of $1 \cdot 55$ cm is dependent on the number and size of the water droplets in the atmospheric column and this fact has been used to deduce precipitation rates. It is interesting that this technique is most suited to ocean areas where the emission of contaminating microwave radiation from the ocean surface is low compared with the emission from land surfaces. These variations in surface emissivity in the microwave spectrum have been used to detect the distribution of sea ice. The higher microwave emissivity of sea ice compared with sea water allows identification of the boundary between sea ice and open water even in cloudy conditions. Daily variations in sea-ice cover can thus be monitored by the satellite network. In addition, it is known that the emissivity of ice more than 1 year old is lower than that of newly formed ice and therefore the age distribution of the ice can be mapped.

In the field of oceanography the development of active microwave satellite instruments is seen as the most promising avenue for obtaining quantitative information about the ocean surface. In 1978 the first oceanographic satellite, SEASAT, was launched. This satellite carried three different types of radar. The first radar, known as the scatterometer, measured the back-scattered radiation from the sea surface at a wavelength of 22 cm. The intensity of the back-scattered radiation is proportional to the surface-wind stress whilst the direction of the scattering is related to the direction of the surface stress. Although the theory of the scattering at this wavelength is not well understood, empirical calculations suggest that this technique can provide accurate wind measurements over the world's ocean. Second, a synthetic aperture radar, SAR, on board SEASAT provided radar images of the sea surface. These images gave information about the surface wave spectrum, and the direction of the waves. Surprisingly, the radar images also detected internal waves in the ocean as well as variations in bottom topography in water which was many tens of metres deep. Since SAR wavelengths are absorbed in the upper centimetre of the ocean, this effect is probably an indirect influence of bottom topography on the surface wave field. The third radar flown on SEASAT was a high-precision radar altimeter which was able to measure the distance between the ocean surface and the satellite to a precision of ∓ 10 cm. Variations in the geometric distance between the sea surface and satellite are related to variations in the satellite orbit, variations in the surface geoid and the ocean surface slope associated with ocean currents. The first effect can be eliminated by accurate position-fixing systems but distinguishing geoid variations from dynamical sea-level variations is very difficult unless independent measurements of the geoid or sea level are available. However, since the geoid is static, repeated passes of the altimeter over a given area can distinguish time-dependent changes in the sea surface elevation with an accuracy of ∓ 10 cm. SEASAT was able to detect time variations of mesoscale eddies in the western Atlantic, as shown in figure 7.9, and it is therefore feasible that, within the next decade, eddy maps of the ocean, similar to the now familiar atmospheric weather maps may become routinely available. In the equatorial ocean sea-level variations are associated with displacement in the shallow equatorial thermocline. High sea levels are generally associated with a deep thermocline and vice versa. Thus it is possible that altimeter measurements may be used to obtain routine observations of the thermocline displacement as well as the heat content in the upper ocean. This technique would be invaluable for measuring seasonal and interannual variations in the equatorial ocean associated with ENSO and other large-scale phenomena. With improvements in the precision of the altimeter, to ∓ 2 cm, and with high-quality determinations of the surface geoid, it would be feasible to determine surface geostropic currents from the satellite measurements, since:

$$v_g(0) = \frac{g}{f}\frac{\partial \xi}{\partial x} + \frac{1}{f\varrho_0}\frac{\partial p_0}{\partial x} \tag{10.1}$$

where $v_g(0)$ is the surface geostrophic current, ξ is the height of the sea surface and p_0 is surface pressure. All other terms have their conventional meaning.

From meteorological surface pressure measurements, the second term in equation 10.1 can be determined and hence v_g at the surface is obtained (figure 10.16). Moreover, from the scatterometer wind-stress data, the wind-drift current, v_e, can be calculated. Thus the total surface current $v_g + v_e$ could be determined.

One major limitation of all satellite techniques is that sea water is opaque to electromagnetic radiation in both the infra-red and microwave regions of the spectrum. Therefore it is not possible to obtain direct information about the ocean below the surface, except perhaps where the deeper ocean influences the surface wave field or the sea level, as discussed above. However, the ocean is transparent to sound and so monitoring of the deeper ocean will depend on developments in acoustic techniques. As already mentioned, neutrally buoyant floats can be monitored in the sound channel over large distances and, in the future, larger hydrophone arrays will be deployed to track these floats. A technique which uses the propagation of sound waves to obtain information on the temperature and velocity structure of the ocean has been recently developed by Munk and Wunch (1983). The technique is known as acoustical tomography and, in principle, it is similar to the use of X-rays in medical physics to scan the brain. The travel time of an acoustic pulse

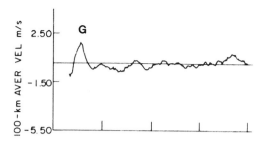

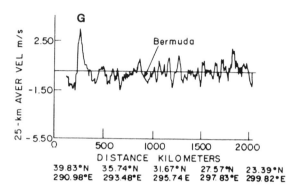

Figure 10.16. Geostrophic velocity determined by satellite altimetry along a track crossing the Gulf Stream and Bermuda. The Gulf Stream, G, and eddy variations to the east of the Gulf Stream can be seen. Two different averaging intervals are displayed. (Reproduced from C. Wunsch and E. M. Gaposchkin (1980), Rev. of Geophysics and Space Physics, **18**, *734, figure 10, copyright of the American Geophysical Union.*

from point A to point B, shown in figure 6.2, is a function of the temperature, salinity and depth of the intervening water. Variations in the water mass between points A and B will produce variations in the travel time of the pulse and hence the changes in the water mass can be monitored. The velocity of sound in sea water is principally controlled by temperature rather than by salinity and so a measure of the heat content of the water mass can be obtained using this technique. Furthermore, because of the transparency of the ocean to sound waves, it is thought that this technique can give measurements on ocean-basin scales although, as yet, it has only been tested over relatively short distances of the order of 100 km. Information on the mean velocity along the line AB can be obtained using co-located sources and receivers. The difference in travel time for an acoustic pulse emitted at point A from that emitted at point B gives the integrated velocity of the water mass between the two points. Average currents can be determined to an accuracy of $\mp 1 \cdot 5$ mm s^{-1} compared with a conventional current meter accuracy of ∓ 1–2 cm s^{-1}.

An advantage of acoustical tomography over conventional techniques is that the number of unique pathways increases geometrically with the number of transceivers and thus a large amount of information can, in principle, be obtained from a relatively modest array.

In conclusion, it can be appreciated that further development of satellite techniques and of acoustical tomography will provide oceanographers and meteorologists with measurements of the ocean and atmosphere on a scale unobtainable from conventional techniques. With ingenuity and foresight, this new type of information should provide oceanographers, meteorologists and climatologists with a more quantitative perspective on the global ocean and atmosphere system.

General reading

Deacon, M. B. 1971, *Scientists and the Sea—a Study of Marine Science. 1650–1900*, 445 pp., (London: Academic Press).

Deacon, M. B. (ed.) 1978, *Oceanography. Concepts and History*, Benchmark Papers in Geology Vol. 35, 394 pp. (London: Dowden, Hutchinson and Ross Inc.).

Defant, A., 1961, *Physical Oceanography*, Vols 1 (729 pp.) and 2 (598 pp.) (Oxford: Pergamon).

Dietrich, G., Kalle, K., Krauss W. and Siedler G., 1980, *General Oceanography: An Introduction*, 626 pp.

Fairbridge, R. W., (ed.), 1966, *The Encyclopedia of Oceanography*, 1021 pp. (New York: Van Nostrand Reinholt).

Gill, A. E., 1982, *Atmosphere–Ocean Dynamics*, 662 pp. (London: Academic Press).

Gross, G. M., 1977, *Oceanography — A View of the Earth*, 497 pp., (Englewood Cliffs, NY: Prentice-Hall).

Harvey, J. G., 1976, *Atmosphere and Ocean:—Our Fluid Environment*, 143 pp. (Artemis Press).

Hess, S. L., 1959, *Introduction to Theoretical Meteorology*, 362 pp. (New York: Holt Reinhart and Winston).

Hill, M. N., 1963, *The Sea: Ideas and Observations, Vol. 1. Physical Oceanography*, 864 pp. (New York: Wiley-Interscience).

Holton, J. R., 1979, *An Introduction to Dynamical Meteorology*, second edition, 391 pp. (New York: Academic Press).

McLellan, H. J., 1965, *Elements of Physical Oceanography*, 151 pp. (Oxford: Pergamon.)

Maury, M. F., 1855, *The Physical Geography of the Sea*, 207 pp.

Monin, A. S. 1972, *Weather Forecasting as a Problem in Physics*, 199 pp. (Cambridge, MA: MIT Press).

Neuman, G. and Pierson, W., 1966, *Principles of Physical Oceanography*, 545 pp. (Englewood Cliffs, NY: Prentice-Hall).

Perry, A. H. and Walker, J. M., 1977, *The Ocean–Atmosphere System*, 160 pp. (London: Longman).

Pickard, G. L. and Emery, W. J., 1982, *Descriptive Physical Oceanography*, fourth edition, 249 pp. (Oxford: Pergamon).

Pond, S. and Pickard, G. L., 1983, *An Introduction to Dynamical Oceanography*, second edition, 329 pp. (Oxford: Pergamon).

Riehl, H., 1979, *Climate and Weather in the Tropics*, 611 pp. (London: Academic Press).

Sverdrup, H. U., Johnson, M. W. and Fleming, R. H., 1942, *The Oceans: Their Physics, Chemistry and General Biology*, 1060 pp. (Englewood Cliffs, NY: Prentice-Hall).

Tchernia, P., 1980, *Descriptive Regional Oceanography*, Pergamon Marine Series, Vol. 3, 253 pp. (Oxford: Pergamon).

Tolmazin, D., 1985, *Elements of Dynamic Oceanography*, 181 pp. (London: Allen and Unwin).

Wallace, J. M. and Hobbs, P. V., 1977, *Atmospheric Science*, 467 pp. (London: Academic Press).

Warren, B. A. and Wunsch, C., 1981, *Evolution of Physical Oceanography*, 623 pp. (Cambridge, MA: MIT Press).

References _____

Chapter 1

Allen, C. W., 1958, Simons Memorial Lecture "Solar Radiation", *Quarterly Journal of the Royal Meteorological Society*, **84**, 307–318.

Goody, R. M. and Walker, J. C. G., 1972, *Atmospheres*, (Englewood Cliffs, NJ: Prentice-Hall).

Henderson-Sellers, A. and Wilson, M. F., 1983, Albedo observations in climatic research, *Philosophical Transactions of the Royal Society of London*, **A309**, 285–294.

Jacobowitz, H., Smith, W. L., Howell, H. B. and Nagle F. W., 1979, The first 18 months of planetary radiation budget measurements from the Nimbus 6 ERB Experiment, *Journal of Atmospheric Science*, **36**, 501–507.

Chapter 2

Brancazio, P. J. and Cameron, A. G. W., 1964, *The Origin and Evolution of Atmospheres and Ocean* (New York: Wiley), 314 pp.

Dittmar, W., 1884, Report on researchs into the composition of ocean-water collected by H.M.S. Challenger during the years 1873–76, in *Report on the Scientific Results of the Voyage of H.M.S. Challenger. Physics and Chemistry*, (London: HMSO), **1**, pp. 199–209, 227–235.

Tolstoy, I. and Clay, C. S., 1966, *Ocean Acoustics: Theory and Experiment in Underwater Sound*, (New York: McGraw-Hill), 293 pp.

Chapter 3

Dopplick, T. G., 1972, Radiative heating of the global atmosphere, *Journal of Atmospheric Science*, **29**, 1278–1294.

Ivanoff, A., 1977, Oceanic absorption of solar energy, in *Modelling and Prediction of the Upper Layers of the Ocean*, ed. Kraus, E. B. (Oxford: Pergamon), Chapter 3.

Jerlov, N. G., 1976, *Marine Optics* (Amsterdam: Elsevier), p. 251.

Kondratyev, K. Ya., 1969, *Radiation in the Atmosphere*, (London: Academic Press), 912 pp.

Liou, K. N., 1980, *An Introduction to Atmospheric Radiation*, (London: Academic Press).

London, J., 1957, *A Study of the Atmospheric Heat Balance: Final Report*, contract AF 19(122)-165, Department of Meteorology, New York University.

Manabe, S. and Strickler, R. F., 1964, Thermal equilibrium of the atmosphere with a convective adjustment, *Journal of Atmospheric Science*, **21**, 361–385.

Paltridge, G. W. and Platt, C. M. R., 1976, *Radiative Processes in Meteorology and Climatology*, 318 pp. (Amsterdam: Elsevier).

Woods, J. D., Barkman, W. and Horch, A., 1984, Solar heating of the oceans: diurnal, seasonal and meridional variation, *Quarterly Journal of the Royal Meteorological Society*, **110**(465), 633–656.

Chapter 4

Ludlam, F. H., 1980, *Clouds and Storms* (University Park, PA: Pennsylvania State University Press), 405 pp.

Mason, B. J., 1975, *Clouds, Rain and Rainmaking*, (Cambridge University Press), 189 pp.

Mason, B. J., 1971, *The Physics of Clouds* (Oxford: Clarendon Press), 671 pp.

Scorer, R. S., 1972, *Clouds of the World* (Newton Abbot, UK: David Charles), 176 pp.

Chapter 5

Broecker, W. S., 1981, Geochemical tracers and ocean circulation, in *Evolution of Physical Oceanography*, ed. Warren, B. A. and Wunsch, C. (Boston, MA: MIT Press), Chapter 15.

Bunker, A. F., 1976, Computations of surface energy flux and annual air-sea interaction cycles in the North Atlantic, *Monthly Weather Review*, **104**, 1122–1140.

Lumb, F. E., 1964, The influence of clouds on hourly amounts of total solar radiation, *Quarterly Journal of the Royal Meteorological Society*, **90**, 43–56.

Oort, A. H. and von der Haar, T. H., 1976, On the observed annual cycle in the ocean–atmosphere heat balance over the northern hemisphere, *Journal of Physical Oceanography*, **6**, 781–800.

Priestley, C. H. B., 1959, *Turbulent Transfer in the Lower Atmosphere* (University of Chicago Press), 130 pp.

Sellers, W. D. 1965, *Physical Climatology* (University of Chicago Press), 272 pp.

Stommel, H., 1980, Asymmetry of interoceanic fresh-water and heat fluxes, *Proceedings of the National Academy of Science, USA*, **77**, 2377–2381.

Stommel, H. and Csanady, G. T., 1980, A relation between the T–S curve and global heat and atmospheric water transports, *Journal of Geophysical Research*, **85**, 495–501.

Worthington, L. V., 1981, Water masses of the world ocean—a fine scale census, in *Evolution of Physical Oceanography*, ed. Warren, B. A. and Wunsch, C. (Boston, MA: MIT Press), Chapter 2.

Chapter 6

Burkov, V. A., Bulatov, R. P. and Neiman, V. G., 1973, Large scale features of water circulation in the world ocean, *Oceanology*, **13**, 325–332.

van der Hoven, J., 1957, Power spectrum of wind velocity in the frequency range from 0.0007 to 900 cycles per hour, *Journal of Meteorology*, **14**, 160–164.

Lorenz, E. N., 1969, *The Global Circulation of the Atmosphere* (London: Royal Meteorological Society).

Luyten, J. R. and Swallow, J. C., 1976, Equatorial undercurrents, *Deep Sea Research*, **23**(10), 999–1001.

Newell, R. E. *et al.*, 1972, *The General Circulation of the Tropical Atmosphere*, (Boston, MA: MIT Press).

Reid, J. L., 1981, On the mid-depth circulation of the world ocean, in *Evolution of Physical Oceanography*, ed. Warren, B. A. and Wunsch, C. (Boston, MA: MIT Press), Chapter 3.

Rossby, H. T., Riser, S. C. and Mariano, A. J., 1983, The western North Atlantic —a Lagrangian viewpoint, in *Eddies in Marine Science*, ed. Robinson, A. R. (New York: Springer-Verlag), pp. 66–88.

Rossby, H. T., Voorhis, A. D. and Webb, D., 1975, A quasi-lagrangian study of mid-ocean variability using long range SOFAR floats, *Journal of Marine Research*, **33**, 355–382.

Swallow, J. C., 1955, A neutral buoyancy float for measuring deep currents, *Deep Sea Research*, **3**, 74–81.

Swallow, J. C., 1971, The Aries current measurements in the western North Atlantic, *Philosophical Transactions of the Royal Society of London*, **A270**, 451–460.

The TWERLE Team, 1977, The TWERL experiment, *Bulletin of the American Meteorological Society*, **58**, 936–948.

Warren, B. A., 1981, Deep circulation in the world ocean, in *Evolution of Physical Oceanography*, ed. Warren, B. A. and Wunsch, C. (Boston, MA: MIT Press), Chapter 1.

Wunsch, C., 1981, Low frequency variability of the sea, in *Evolution of Physical Oceanography*, ed. Warren, B. A. and Wunsch, C. (Boston, MA: MIT Press), Chapter 11.

Chapter 7

Cheyney, R. E. and Marsh, J. G., 1981, Seasat altimeter observations of dynamic topography in the Gulf Stream Region, *Journal of Geophysical Research*, *C1*, **86**, 473–483.

Findlater, J., 1971, *Geophysical Memoirs*, No. 115, 53 pp. (London: HMSO).

McWilliams, J. C., 1976, Maps from the mid-ocean dynamics experiment. Part 1. Geostrophic streamfunction, *Journal of Physical Oceanography*, **6**, 810–827.

Stommel, H., 1958, The abyssal circulation, *Deep Sea Research*, **5**, 80–82.

Stommel, H., 1966, *The Gulf Stream* (Cambridge University Press/University of California Press), 248 pp.

Stommel, H. and Arons, A. B., 1960, On the abyssal circulation of the world's ocean. 1. Stationary planetary flow patterns on a sphere, *Deep Sea Research*, **6**, 140–154.

Swallow, J. C. and Bruce, J. G., 1966, Current measurements off the Somali coast during the SW monsoon of 1964, *Deep Sea Research*, **13**, 861–888.

Chapter 8

Cartwright, D. E., Driver, J. S. and Tranter, J. E., 1977, Swell waves at Saint Helena related to distant storms, *Quarterly Journal of the Royal Meteorological Society*, **103** (438), 655–684.

Charnock, H., 1981, Air—sea interaction, in *Evolution of Physical Oceanography*, ed. Warren, B. A. and Wunsch, C. (Boston, MA: MIT Press), Chapter 17.

Flather, R. A., 1984, A numerical model investigation of the storm surge of 31 January and 1 February 1953 in the North Sea, *Quarterly Journal of the Royal Meteorological Society*, Vol 110(465), 591–612.

Kinsman, B., 1984, *Wind Waves: Their Generation and Propagation on the Ocean Surface* (New York: Dover Publications), 676 pp.

Pekeris, C. L. and Accad, Y., 1969, Solution of Laplaces equations for the M_2 tide in the world's ocean, *Philosophical Transactions of the Royal Society of London*, **A265**, 413–436.

Revault-d'Allonnes, M. and Caulliez, G., 1980, Observations of air-sea interaction parameters in the open ocean, in *Marine Turbulence*, ed. Nihoul, J. C. J. (Amsterdam: Elsevier), pp. 103–121.

Thomson, W. (Lord Kelvin), 1879, On gravitational oscillations of rotating water, *Proceedings of the Royal Society, Edinburgh*, **10**, 92–100.

Chapter 9

Lorenz, E. N., 1967, *The Nature and Theory of the General Circulation of the Atmosphere* (World Meteorological Organization), 161 pp.

Newell, R. E., Vincent, D. G., Dopplick, T. G., Ferruzza, D. and Kidson, J. W., 1969, The energy balance of the global atmosphere, in *The Global Circulation of the Atmosphere*, ed. Corby, G. A. (London: Royal Meteorological Society), pp. 42–90.

Newton, C. W., 1969, The role of extratropical disturbances in the global atmosphere, in *The Global Circulation of the Atmosphere*, ed. Corby, G. A. (London: Royal Meteorological Society), pp. 137–158.

Palmen, E. and Newton, C. W., 1969, *Atmospheric Circulation Systems* (London: Academic Press), 603 pp.

Parker, C. E., 1971, Gulf Stream rings in the Sargasso Sea, *Deep Sea Research*, **18**(10), 981–993.

Chapter 10

Cane, M. A., 1983, Oceanographic events during El Nino, *Science*, **222**, 1195–1202.

Gates, W. L., 1976, Modelling the ice-age climate, *Science*, **191**, 1138–1144.

Ghil, M., 1980, Internal climatic mechanisms participating in glaciation cycles, in *Climatic Variations and Variability: Facts and Theories*, ed. Berger, A., NATO Advanced Study Series (Dordrecht: Reidel), pp. 539–558.

Imbrie, J., 1980, Time dependent models of the climate response to orbital variations, in *Climatic Variations and Variability: Facts and Theories*, ed. Berger, A., NATO Advanced Study Series (Dordrecht: Reidel).

JASIN, 1983, Results of the Royal Society Joint Air–Sea Interaction Project (JASIN), *Philosophical Transactions of the Royal Society of London*, **A308**, 221–449.

Mitchell, J. M., A note on solar variability and volcanic variability, in *Physical Basis of Climate and Climate Modelling* (World Meteorological Organization), Garp 16.

Munk, W. and Wunsch, C., 1983, Observing the ocean in the 1990's, *Philosophical Transactions of the Royal Society of London*, **A307**, 438–463.

Munk, W. and Wunsch, C., 1983, The study of the ocean and the land surface from satellites, *Philosophical Transactions of the Royal Society of London*, **A309**, 243–464.

Namias, J., 1971, The 1968–69 winter as an outgrowth of sea and air coupling during antecedent seasons, *Journal of Physical Oceanography*, **1**, 65–81.

Namias, J., 1973, Hurricane Agnes—an event shaped by large scale air–sea systems generated during antecedent months, *Quarterly Journal of the Royal Meteorological Society*, **99**(421), 506–519.

Niehaus, F., 1980, The impact of energy production on atmospheric carbon dioxide concentrations, in *Climatic Variations and Variability: Facts and Theories*, ed. Berger, A., NATO Advanced Study Series (Dordrecht: Reidel), pp. 641–660.

Pittock, A. B., Frakes, L. A., Jenssen, D., Peterson, J. A. and Zillman, J. W. (eds), 1978, *Climate Change and Variability: A Southern Perspective* (Cambridge University Press), 455 pp.

Smagorinsky, J., 1953, The dynamical influence of large scale heat sources and sinks on quasi-stationary mean motions of the atmosphere, *Quarterly Journal of the Royal Meteorological Society*, **79**(341), 342–366.

WMO, Garp Publication Series 16, 1975, *The Physical Basis of Climate and Climate Modelling*.

WMO, Garp Publication Series 24, 1981, *Scientific Results of the Air Mass Transformation Experiment*, 236 pp.

Index